Deltas of the World

Series Editor

Abdelazim Negm, Zagazig, Egypt

Worldwide, over 600 million people live in deltas. Extensive human activities in deltas, their supplying rivers, and related maritime currents increase the pressure on deltas' natural resources and habitat equilibrium, leading to degradation of the available resources and jeopardizing inhabitants' livelihoods – and in some cases even their lives.

This book series is intended to promote a sustainable future for deltas around the globe and the communities that live in them. Accordingly, we publish studies focusing on various aspects of deltas, including hydrogeology, soil pollution, water pollution, environmental hydrology, environmental hydraulics, groundwater engineering and management, morphology, anthropology, socioeconomics, wastewater management, water resources engineering and management, agriculture, greenhouse agriculture, irrigation and drainage engineering, hydraulic structures, geophysics and geology, sedimentology, mineralogy, remote sensing and GIS studies, climate change variabilities and impacts, marine-delta interactions and sustainability, and solid waste management. The findings presented here support the SDGs and Agenda 2030 with regard to helping countries with deltas and related environments to preserve their natural resources for future generations.

The editors invite scientists, researchers and scientific communities to contribute to the series by submitting delta atlases, edited books, monographs, and high-quality conference proceedings focused on deltas around the world.

Subrata Mondal · Rupak K. Paul

Climate Variability and Human Interference

Hydrologic Transformation of the South Koel River Basin and Linkages to the Downstream Deltaic System, Eastern India

Springer

Subrata Mondal
Department of Geography
Acharya Brojendra Nath Seal College
Cooch Behar, West Bengal, India

Rupak K. Paul
Department of Geography and Applied
Geography
University of North Bengal
Siliguri, West Bengal, India

ISSN 2731-832X ISSN 2731-8338 (electronic)
Deltas of the World
ISBN 978-3-032-13256-7 ISBN 978-3-032-13257-4 (eBook)
https://doi.org/10.1007/978-3-032-13257-4

This Springer imprint is published by the registered company Springer Nature Switzerland AG
The registered company address is: Gewerbestrasse 11, 6330 Cham, Switzerland

Dedicated to all well-wishers

Preface

This book presents a comprehensive investigation into the hydrologic transformation of the South Koel River Basin in Eastern India, emphasizing the dynamic interactions between climate variability and human activities. It further explores the linkages between the basin and the downstream Brahmani-Baitarani deltaic system.

The South Koel River originates near Nagri village (Ranchi District, Jharkhand) at about 630 m elevation, then flows eastward through the Chotanagpur Plateau until it meets the Sankh River at Vedvyas near Rourkela (Odisha). There, their combined flow becomes the Brahmani River, which continues southeast and ultimately contributes to the Brahmani–Baitarani deltaic system. Although the South Koel River basin is non-deltaic, changes in its hydrologic regime (discharge, sediment dynamics, and nutrient supply) have cascading effects on downstream systems. These upstream changes shape the Brahmani River's flow, sediment regime, and ecological health, and ultimately influence the Brahmani–Baitarani deltaic system. Thus, this book examines the basin in a broader fluvial context, emphasizing its indirect but significant connectivity to the deltaic environment.

As a monsoon-driven river, the South Koel represents many tropical basins which are undergoing rapid hydrological changes caused by shifting climatic conditions and intensifying human activities. This study seeks to disentangle these forces using a combination of long-term statistical trend analysis, hydrological modeling, and land use sensitivity assessments.

The study begins by laying a conceptual and thematic foundation in the introductory chapter (Chap. 1), which contextualizes the relevance of examining the hydrologic impacts of climate change and anthropogenic intervention. It represents a synthesis of global and regional debates on runoff variability, establishing the broader significance of basin-scale studies in vulnerable regions like Eastern India. The narrative progresses into a critical review of existing literature, where Chap. 2 presents a structured and detailed assessment of past scholarship, methods, and emerging knowledge gaps. This review sets the groundwork for formulating the study's specific objectives and guiding questions.

Chapter 3 introduces the South Koel River Basin in detail, outlining its geographic, geologic, climatic, and socio-economic attributes. This chapter also presents the methodological framework of the study, including the datasets used, the model architecture (SWAT model), and the analytical techniques applied. These foundational elements support the technical depth of the analysis that follows.

In Chap. 4, the book undertakes a time-series investigation of rainfall and runoff data to detect gradual trends and abrupt shifts over time. Techniques such as the Buishand U Test, Buishand Range Test, Standard Normal Homogeneity Test, Pettitt's Test, Mann-Kendall Trend Test, Modified Mann Kendall Test, and Sequential Mann-Kendall Test are used to evaluate the statistical significance of changes across temporal scales. This chapter offers a historical hydrological baseline for subsequent analyses.

Building upon this, Chap. 5 focuses on quantifying the impacts of climate variability and human activity on runoff. By calibrating and running multiple simulation scenarios in SWAT, the chapter isolates the effect of meteorological variations from land use and land cover changes, revealing the relative strength of each driver in altering runoff patterns.

Chapter 6 deepens the analysis by evaluating runoff sensitivity to different land use and land cover types. Forest cover, agricultural expansion, and urban built-up areas are examined through HRU-based categorization and statistical models to identify threshold effects and spatial dependencies. This analysis reveals not only how much runoff changes, but also under what land use configurations such changes become hydrologically significant.

Chapter 7 examines the upstream–downstream connectivity between the South Koel Basin and the Brahmani-Baitarani delta, emphasizing how basin-scale hydrologic changes-driven by climate variability and human interventions-affect deltaic flow, sediment delivery, and ecological health. Drawing on global and regional case studies, it highlights the cascading effects of upstream alterations on delta resilience and underscores the need for integrated basin-delta management to address environmental flows, sediment continuity, and climate-induced vulnerabilities.

In Chap. 8, the major findings of the study are synthesized and interpreted in the broader context of water security, catchment resilience, and landscape sensitivity. This discussion provides insight into basin-specific vulnerabilities and the hydrologic implications of ongoing land use and land cover changes.

The final chapter (Chap. 9) concludes the book with targeted policy recommendations and suggestions for future research. Emphasis is placed on sustainable land and water management strategies tailored to monsoon-fed basins, highlighting the need for integrated planning approaches that balance environmental constraints with developmental goals.

All references and citations of this book follow the American Psychological Association (APA) style, sixth edition, and American spelling conventions. References are provided at the end of each chapter for ease of access and contextual clarity.

This book is intended for geographers, environmental scientists, hydrologists, and water resource managers, as well as policy-makers and postgraduate students who seek to understand the intricacies of how climate and human activity influence runoff. We hope that the results and methodologies outlined here not only enrich the academic discourse but also inform future efforts toward sustainable and adaptive basin-scale management.

Cooch Behar, India Subrata Mondal
Siliguri, India Rupak K. Paul

Competing Interests The authors have no competing interests to declare that are relevant to the content of this manuscript.

Contents

Acronyms and Abbreviations

ANN	Artificial Neural Network
C	Centigrade
cumec	Cubic meter per second
CWC	Central Water Commission
DEM	Digital Elevation Model
GLUE	Generalized Likelihood Uncertainty Estimation
GSI	Geological Survey of India
HYV	High Yielding Variety
IMD	India Meteorological Department
km	Kilometer
km^2	Square Kilometer
LULC	Land use and land cover
m	Meter
MCM	Million Cubic Meter
MCMC	Markov Chain Monte Carlo
mm	Millimeter
n	Number
NASA	National Aeronautics and Space Administrations
NSE	Nash Sutcliff Efficiency
OLI	Operational Land Imager
ParaSol	Parameter Solution
PSO	Particle Swarm Optimization
PTF	Pedo-transfer Function
SCS-CN	Soil Conservation Services Curve Number
SPSS	Statistical Package for Social Sciences
SRTM	Shuttle Radar Topographic Mission
SUFI-2	Sequential Uncertainty Fitting Version 2
SWAT	Soil and Water Assessment Tool

TM	Thematic Mapper
USGS	United States Geological Survey
WGS84	World Geodetic System 1984
yr	Year

List of Figures

List of Tables

List of Equations

Introduction 1

1.1 Prelude

Runoff, a fundamental component of the hydrological cycle, is the portion of precipitation that flows out of a river basin after satisfying losses through evapotranspiration and various storage forms such as interception and surface detention. It is measured as runoff rate (discharge), the flow per unit time (cumec or cusec); runoff volume, the total yield over a specified period (m^3, km^3, BCM, acre-feet, or hectare-meters); and runoff depth, the volume normalized by basin area (mm or cm), making it directly comparable with rainfall depth.

Runoff dynamics further describe the patterns and variations of runoff over time or across space, offering insight into the hydrological behavior of a basin. As a fundamental indicator of water availability, runoff dynamics are profoundly influenced by the simultaneous effects of climate variability and human activities - the two primary drivers of runoff alteration at the basin scale. At larger spatial scales (e.g., continental or hemispheric), however, these drivers become functionally interdependent (Seyoum et al. 2015; Wang et al. 2012), jointly modifying catchment hydrology (Giambelluca 2005). Such alterations have significant implications for water supply, floods and droughts, erosion and sedimentation, ecosystem health, and human well-being, irrespective of spatial or temporal extent.

Runoff alteration/change in a river basin is understood usually from two perspectives: *gradual change* (also called *trend*), and *step change* (also called *sudden* or *abrupt change*). Understanding these processes, particularly in monsoon-driven tropical basins like the South Koel, is critical for water resource planning since enormous pressure is exerted on freshwater resources through these processes. Moreover, such changes have a profound downstream impact on deltaic systems such as the Brahmani–Baitarani.

S. Mondal and R. K. Paul, *Climate Variability and Human Interference*, Deltas of the World, https://doi.org/10.1007/978-3-032-13257-4_1

Direct human intervention in the basin hydrologic system like water abstraction for domestic water use, irrigation, and agricultural practices, industrialization (Liu et al. 2019) and water diversion (Wang et al. 2016) etc., and ensuing runoff change can be easily discernible and accordingly, the remedial measures can be formulated, but indirect human interference through land use and land cover (LULC) change often remains invisible (Afonso de Oliveira Serrão et al. 2022). That is why the issue does not receive priority or importance in water-resources planning. But this LULC change coupled with climate variability can irreversibly alter basin-scale hydrology, with adverse impacts extending into future generation (Rivas-Tabares et al. 2022).

Although it is a global phenomenon, the issue has regional implications because of significant geographical variations in river basin characteristics across the continents, and hence, the runoff response is getting manifested differently around the globe (Vörösmarty et al. 2000) depending on the river basin scale (Anil and Ramesh 2017; Khoi et al. 2019; Pirnia et al. 2019; Wang et al. 2020; Xu et al. 2014b), degree of human interference in the river basin (Chen et al. 2005; Liang et al. 2015a; Tang et al. 2021; Zhang et al. 2020), variation of geophysical condition of river basin (Khoi et al. 2019; Ma et al. 2009; Zeng et al. 2015; Zhou et al. 2020), climate types (Li et al. 2010; Wang et al. 2017, 2021; Zhang et al. 2017) and so on.

The general scenario that emerges from previous studies (Wang et al. 2012; Anil and Ramesh 2017) is that indirect human interventions—primarily through LULC change—may alter runoff in either direction; i.e., they may increase or decrease runoff depending on the nature of the intervention. By contrast, direct interventions such as dams, reservoirs, flow diversions, and water withdrawals for domestic or other uses generally lead to a reduction in streamflow. The impact of climate change, on the other hand, is not yet fully explored.

Moreover, there is no consensus on whether climate variability or human activities exert the dominant influence on runoff alteration. Swain and Patra (2019) studied 32 ungauged catchments of tropical monsoon climate in India spreading over Jharkhand, West Bengal, Chhattisgarh, Orissa, Telangana, Andhra Pradesh, Tamil Nadu, Karnataka, and Maharashtra, and their findings suggested that the role of climate variability is greater than the human activities to reduce the runoff during 1990–2011. Musie et al. (2020), on the other hand, reported the exact opposite result for Ketar and Meki watersheds in Ethiopia.

Since the process is not straightforward, a region-specific investigation is needed to explore the functional relationship between river response and LULC changes amidst the changing climate. Otherwise, it would be very difficult to formulate a region-specific effective remedial strategy.

In this study, an attempt has been made to address these issues through a focused investigation of the South Koel River Basin in Eastern India, providing empirical insights into runoff alteration mechanisms and their broader linkages to downstream deltaic systems.

1.2 Statement of the Research Problem

The present work examines runoff alteration in the South Koel River Basin under a tropical monsoon climate, focusing on the influences of climate variability and human activities. It seeks to quantify and evaluate changes in runoff generation mechanisms that evolve over time through the complex interactions between climatic variations and human activities.

Here, the South Koel River basin has been chosen as the study area because of its diverse surface characteristics, as reported in many literature (CWC 2014; Tirkey et al. 2018). The basin extends over the plateau and plain region, and hence there is diversity in soil types that lead to diverse land use and land cover (LULC), particularly agricultural practices. Substantial changes have occurred in climate and LULC over the past couple of decades, and that is why the area seems to be ideal for addressing the runoff change mechanism with the full extent of its complexity.

As reported by several researchers (Tang et al. 2013; Uniyal et al. 2015; Wu et al. 2019; Yuan et al. 2019), the response of basin hydrologic system varies across the regions depending upon geographical conditions like climate, topography, soil, human culture and many other factors and even, basins with similar geophysical conditions produce different response depending upon the extent of the basin. Similar catchment responses were also found by many researchers (Li et al. 2019; Afonso de Oliveira Serrão et al. 2022) when they explored only the role of human activities in the alteration of runoff. So, it indicates that the runoff generation mechanism has become more complex under the condition of climate change.

There is also no uniformity across the spaces in the magnitude of the contribution of climate variability and human activity to cause a change in system response due to having non-linearity in the functional relationship between these two, which makes the system response more complex (Ashu and Lee 2020; Ercan et al. 2020; Mahmoodi et al. 2021).

The research problem as stated here has already been explored in many parts of the globe, but no attempt has been made to investigate it under the diverse geophysical conditions of the South Koel River Basin. It would not be rational to infer those findings to the present study area since the system dynamics, as mentioned earlier, differs across the geophysical environments. A few studies (Chaudhary and Pandey 2022; Pandey 2017; Stuti et al. 2020; Swain and Patra 2019), on the other hand, have been carried out in the South Koel River basin, but they have not addressed the present problem. So, the response pattern of streamflow to climate variability and human activity is really an unknown aspect of the hydrological system in the context of the South Koel River basin that has created a research gap, and this research effort is an attempt to fulfill the same by comprehensively examining runoff alterations and situating their implications within the broader Brahmani–Baitarani deltaic context.

1.3 Research Questions

To answer these critical knowledge gaps, this study posed the following key research questions.

(i) What types of changes have occurred in the stream flow: gradual change (trend) or sudden change (step change) or both?
(ii) How much change in runoff is due to climate variability and how much is due to human activities?
(iii) What is the threshold change value in forest cover, agricultural land, and urban built-up area that has functional impact on runoff alteration?
(iv) How do upstream hydrologic transformations in the South Koel Basin influence downstream deltaic systems, particularly the Brahmani–Baitarani delta?

1.4 Objectives

This book was carried out to fulfill the following four objectives:

(i) To assess the temporal variability of rainfall and runoff since 1973 in terms of trend and step change in South Koel River Basin.
(ii) To determine the relative contribution of climate variability and human activities to cause the alteration of runoff in the study area.
(iii) To capture the runoff sensitivity to changes in specific LULC classes with special reference to forest cover, agricultural land use, and urban built-up area in the concerned river basin.
(iv) To analyze how upstream hydrologic transformations in the South Koel Basin influence downstream deltaic systems, with emphasis on the Brahmani–Baitarani delta.

1.5 Methodology in Brief

The present study adopts an integrated methodological framework combining remote sensing, GIS-based spatial analysis, hydrological modeling, and statistical techniques to investigate runoff alterations in the South Koel River Basin. The Soil and Water Assessment Tool (SWAT) model forms the core of the hydrological simulation, calibrated and validated with observed discharge data to ensure reliability.

Time-series statistical methods—including the Buishand U Test, Buishand Range Test, Standard Normal Homogeneity Test, Pettitt Test, Mann–Kendall Trend Test, Modified

Mann Kendall Test, and Sequential Mann–Kendall Test—are applied to detect both gradual trends and abrupt shifts in rainfall and runoff over temporal scales. These analyses provide a baseline for evaluating hydrologic variability.

To disentangle the drivers of runoff change, scenario-based simulations are used to isolate the relative contributions of climate variability and land use/land cover (LULC) changes. Further, forest cover, agricultural land, and urban built-up area impacts are quantified using one-way ANCOVA model.

Finally, this study incorporates upstream–downstream hydrologic connectivity, linking basin-scale runoff dynamics to downstream Brahmani–Baitarani deltaic systems. By combining hydological modeling, statistical techniques, and spatial analysis, the methodology provides a robust framework to examine climate–land use interactions, hydrologic alterations, and their implications for basin and delta resilience.

1.6 Scope and Limitation

It is expected that the findings and the database produced in current research would be very useful input for formulating a sustainable water resource management plan in the South Koel River basin. At the same time, the study is likely to have some limitations due to the lack of a proper database on many soil and rock parameters. For example, saturated hydraulic conductivity data at regular spatial intervals is essential to get accurate simulation results, but these data are not available. To meet up the present requirement, this kind of parameter was estimated using pedo-transfer functions (PTF) but observed data gives the best result, which is beyond the scope of the present study. Soil data, on the other hand, has a spatial resolution of 250 m which is also a bit coarser for such a basin where topographic diversity is more, and thus ensuing soil diversity. Furthermore, there is a lack of direct observational data from the delta region, which constrains the validation of upstream–downstream hydrologic linkages and their ecological impacts. However, the methodology is robust enough to produce the expected result that can contribute to addressing the water-related issue.

1.7 Significance of the Study

Understanding the influence of both climate variability and anthropogenic factors on runoff is a pressing concern for hydrologists, environmental planners, and policy-makers, especially in monsoon-dependent tropical regions. The South Koel River Basin, with its geo-physically diverse and socio-economically important setting, serves as an ideal case to reveal the complex interplay between climatic variability and land-use dynamics.

The present work contributes to the broader discourse on hydrological transformation by combining empirical modeling with spatial analysis, offering not only theoretical insights but also practical implications for water resource planning. It addresses a significant research gap in Eastern India and sets a precedent for similar studies in geo-physiographically complex regions.

By linking basin-scale hydrologic alterations with their downstream implications for deltaic system, this study underscores the importance of upstream–downstream connectivity in sustainable water governance. Its findings provide a decision-support framework for designing integrated land and water management strategies applicable to tropical monsoon-fed river systems worldwide.

1.8 Structure of the Book

This volume is organized into eight chapters, providing a logical and comprehensive exploration of the hydrological transformation in the South Koel River Basin and its linkages to the downstream deltaic system.

This chapter introduces the research problem, objectives, and contextual background.

Chapter 2 presents a global and regional review of existing literature, identifying research gaps and the need for localized analysis.

Chapter 3 describes the study area in detail, outlines data sources, and explains the methodological framework including the SWAT model and statistical tools used.

Chapter 4 analyzes trends and step changes in rainfall and runoff over time using statistical tests such as Buishand U Test, Buishand Range Test, Standard Normal Homogeneity Test, Pettitt Test, Mann–Kendall Trend Test, Modified Mann Kendall Test, and Sequential Mann-Kendall Test.

Chapter 5 quantifies the relative contributions of climate variability and land use and land cover changes to runoff alterations through SWAT simulation-based approaches.

Chapter 6 explores the sensitivity of runoff to specific LULC changes, focusing on forest degradation, agricultural expansion, and urbanization.

Chapter 7 discusses how upstream hydrologic transformations can influence the downstream delta region.

Chapter 8 synthesizes the major findings, providing scientific insights into runoff response mechanisms and regional vulnerabilities.

Chapter 9 concludes with policy recommendations and directions for future research and water governance in monsoon-driven river basins.

1.9 Concluding Remarks

This chapter has laid the foundation of the present research by framing the issue of runoff alteration in the South Koel River Basin under the influences of climate variability and human activities. It has articulated the research problem, questions, and objectives, while also presenting a brief methodological overview, the scope and limitations of the study, and its broader significance. Collectively, these elements provide a coherent groundwork for the chapters that follow, which build upon this framework through a critical review of existing literature and empirical analyses aimed at disentangling the complex dynamics of runoff alteration and its upstream–downstream linkages with the deltaic environment.

State-of-the-Art Review

2

2.1 Prelude

A comprehensive review of the existing literature provides the conceptual and theoretical basis for this study. The present review is structured around two key dimensions of basin-scale hydrological processes. The first dimension encompasses the influence of climate variability and human activities on runoff alteration, drawing insights from global to basin-scale perspectives. The second one incorporates upstream–downstream hydrologic connectivity, with particular attention to how basin-scale transformations affect deltaic systems, focusing on the Brahmani–Baitarani delta in Eastern India. The ultimate objective of examining past works within these two dimensions is to identify critical knowledge gaps, which in turn guide the scope and direction of the present study.

In addition, the review assesses the strengths and limitations of existing methodologies and evaluates the consistency of findings, aiming to develop a deeper understanding of the complex interactions among hydrological factors that shape runoff dynamics. The chapter concludes by anchoring the present study with this scholarly landscape, underlining the need for a region-specific assessment of runoff alteration in the context of South Koel River Basin. This forms the foundation for addressing the central research questions of the study and situating its contributions in the broader discourse of climate-human-hydrology interactions.

S. Mondal and R. K. Paul, *Climate Variability and Human Interference*, Deltas of the World, https://doi.org/10.1007/978-3-032-13257-4_2

2.2 Review of the Impacts of Climate Variability and Human Activities on Runoff Alteration

2.2.1 Selection and Categorisation of Literature

To establish a robust theoretical foundation for this study, a systematic literature review (SLR) was conducted, focusing on research addressing the impacts of climate variability and land use/land cover (LULC) change on runoff. The review followed a structured process involving formulation of guiding questions, systematic database searches, multi-stage screening, and thematic categorization (Table 2.1).

Research articles were retrieved from Google Scholar, Scopus, and the SWAT literature database, using predefined search strings such as "Runoff climate human activities", "Runoff change climate variability human activities," and "Runoff alteration climate human activities" (Table 2.1). Both reference chaining and forward citation tracking were applied to identify additional relevant studies.

A total of 536 studies were collected, of which 110 papers pertaining to the assessment of the impact of climate variability and human activities on runoff alteration were selected for full-text evaluation after a thorough screening of the titles of the literature followed by abstracts, findings, and full-text assessment. Additional literature were collected from the search of the reference list ($n = 19$), citation list ($n = 11$), and authors' personal desk ($n = 9$) (Table 2.1).

The screened literature (149 pieces) was then organized thematically, examining factors such as spatial scale, climate type, degree of human interference, physiography, and analytical techniques. This structured categorization ensured that global, regional, and basin-specific contexts were systematically compared, thereby positioning the South Koel River Basin study within the broader hydrological research landscape.

By synthesizing these findings, this review not only consolidates current understanding of runoff alteration but also lays the groundwork for identifying critical research gaps, particularly those related to tropical monsoon-fed basins and their upstream–downstream linkages.

The basins selected for runoff alteration studies worldwide have been classified into nine categories according to basin size, following the framework of the All India Soil and Land Use Survey (1990) and Singh (1994). The review reveals that the majority of research has been conducted at the basin scale (20,000–300,000 km^2), which accounts for 34.16% of the total studies. This is followed by the sub-catchment scale (2000–10,000 km^2) with 24.22%, the catchment scale (10,000–20,000 km^2) with 16.77%, the watershed scale (500–2000 km^2) with 13.66%, and the regional scale (>300,000 km^2) with 8.70%. By contrast, only a very limited number of studies (2.49%) have investigated hydrological responses at finer spatial units, namely the mini-watershed (<1 km^2), micro-watershed (1–10 km^2), mili-watershed (10–100 km^2), and sub-watershed (100–500 km^2) levels. These smaller-scale studies remain underrepresented despite their potential to provide valuable

Table 2.1 Process followed in the systematic literature review (SLR)

Sl. No.	Name of the process	Details
1.	Formulation of the questions	a. How does the functional response of runoff change in relation to the spatial scale of the area under study? b. How does the degree of human interference affect the variability of runoff responses? c. What are the variations in the functional response of runoff due to changes in climate regions? d. Does the variation in elevation affect the responses of runoff? e. How has the impact on runoff alteration been analyzed?
2.	Name of the search string	"Runoff climate human activities" "Runoff change climate variability human activities" "Runoff alteration climate human activities" "Runoff variability climate change human activities" "Runoff climate land use land cover" "Runoff climate forest" "Stream flow climate human activities" "Discharge climate human activities" "Runoff climate variability human activities" "Runoff climate human activities India" "Runoff climate land use and land cover India" "Runoff climate land use and land cover Jharkhand" "Runoff climate human activities South Koel"
3.	Screening of the collected literature (536)	Literature excluded based on the title screening, n = 216 Literature excluded based on the abstract, n = 147 Literature excluded based on the findings, n = 28 Excluded after full-text evaluation, n = 35 Eligible literature, n = 110
4.	Literature included in the present study (149)	Full-text evaluation, n = 110 Searching reference list, n = 19 Searching citation list, n = 11 Authors' personal desk, n = 9

Source Authors, based on selected literature

insights into localized runoff alteration processes under climate variability and human intervention (Fig. 2.1).

The issues were extensively explored in semi-arid (21.66%) and sub-tropical climatic regions (21.02%), followed by temperate climates (15.92%), areas transitioning from semi-arid to arid climates (11.46%), monsoon climates (9.55%), tropical climates (6.37%), transitional zones from semi-humid to sub-arid climates and arid climates (4.46% each), and Mediterranean climates (3.18%) respectively (Table 2.2).

So far as the dominant LULC is considered, it is observed that most of the works were carried out in agricultural watersheds (60.49%), followed by mixed watersheds

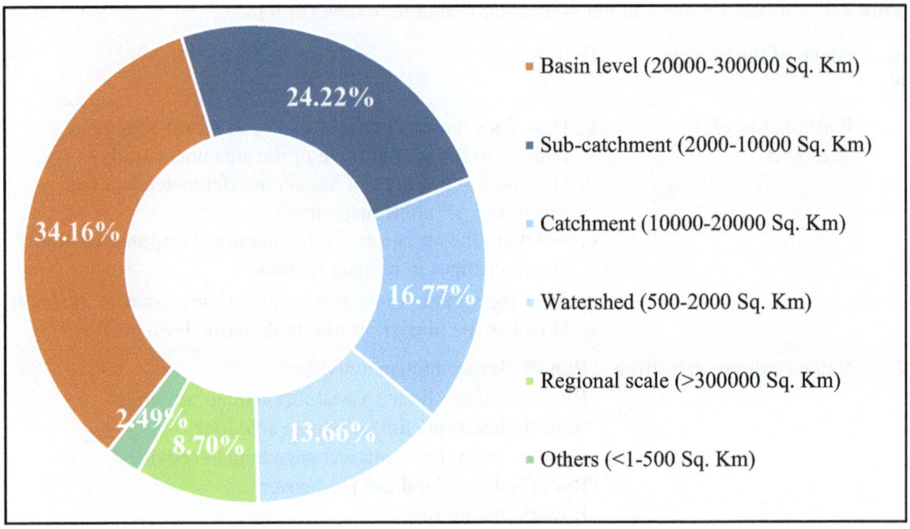

Fig. 2.1 Literature classified based on the spatial scale of the study area reported in literature (After All India Soil and Land Use Survey 1990; Singh 1994)

Table 2.2 Studies performed in different climatic regions

Type	Number of journal articles	Percentage
Tropical climate	10	6.37
Sub-tropical climate	31	21.02
Semi-humid to sub-arid climate	7	4.46
Semi-arid climate	32	21.66
Semi-arid to arid climate	17	11.46
Arid climate	7	4.46
Monsoon climate	14	9.55
Mediterranean climate	5	3.18
Temperate climate	25	15.92
Sub-arctic	3	1.91

Source Authors

(25.31%), forested watersheds (7.41%), grassland watersheds (6.17%), and urban watersheds (0.62%). However, runoff responses have not yet been investigated in coastal, wetland/marsh, and desert watersheds (Fig. 2.2).

On the other hand, a significant amount of research has been conducted in mainly two physiographic regions, namely plain plateau complexes and plateau regions. Nonetheless,

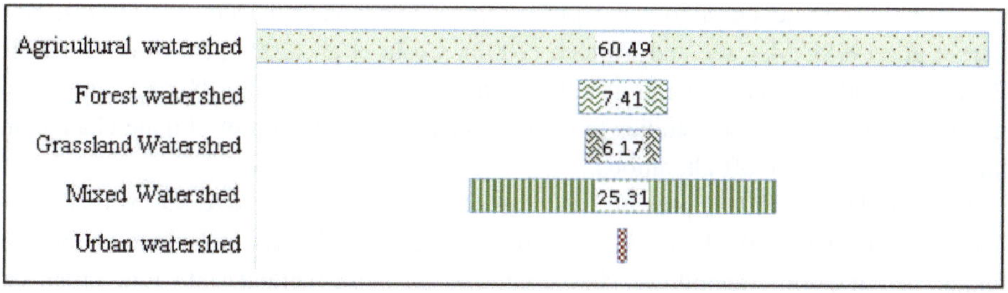

Fig. 2.2 Classification of basin based on dominant land use and land cover as reported in literature

there is still insufficient research regarding alterations in runoff within other climatic regions.

2.2.2 Methodological Review

The study of runoff response to changes in geophysical conditions dates back to the beginning of the twentieth century when the investigation began using paired catchment approaches. Pioneering researchers in that field (Engler 1919; Bates and Henry 1928; Hoyt and Troxell 1934; Hoover 1944; Dunford and Fletcher 1947; Langbein and Wrather 1949) were successful in deriving that forest cover is inversely proportional to basin runoff.

The major limitation of this early approach is that only the hydrologic impact of vegetation cover and settlement in some cases could be assessed. Secondly, the approach could not be applied to medium and large catchments. Since the 1960s, the paired catchment approach started to be superseded by empirical studies based on statistical models, particularly multiple regression, to find out the causal relationship among hydrological variables. The pioneers in this field are Eschner and Satterlund (1966), Emery (1966), Hibbert (1967), and many more. Their works obviously validated the previous findings, and the relative contribution of individual causal and controlling variables was possible to be captured in their attempt.

From the 1970s and 1980s onward, researchers started paying much attention to formulating physical-based models to explore the exact functional relationship of the catchment variables in the framework of the hydrological system. It became possible to address many applied hydrological issues related to river basin management under the condition of climate change (e.g., Karl and Riebsame 1989; Bultot et al. 1990; Nash and Gleick 1991).

The runoff prediction by using a mathematical model then became the major area of investigation in order to estimate and assess the availability of fresh water for future use. Karunanithi et al. (1994), Lorrai and Sechi (1995), and Zealand et al. (1999) suggested that the ANN can be used as a good predictor model for rainfall-runoff analysis.

Presently, researchers across the globe, e.g., Changnon and Demissie (1996), Kristian et al. (1998), McCabe and Wolock (2002), Costa et al. (2003), Jones et al. (2006), Poelmans et al. (2011), Liang et al. (2015) and many more, are mostly concentrating on mathematical model-based studies to capture relative contribution of climate change and human activity on runoff alterations.

Dey and Mishra (2017) provided a consolidated methodological review which identified seven distinct methods for disentangling the relative impacts of climate change and human activities on streamflow. They organized all these methods into four broad categories: experimental approaches (time trend analysis and paired catchment method), hydrological modeling approaches, conceptual approaches (Budyko-based elasticity/decomposition and the Tomer–Schilling framework), and analytical approaches (climate elasticity and hydrological sensitivity methods). Later, Zeng et al. (2020) presented a more comprehensive framework in which the available techniques/approaches were classified into six broader categories, namely hydrological models, scenario combination, Budyko framework, paired catchments, empirical statistical methods, and machine learning. Based on the key insights from these two major review studies, the available methods can be further categorized into six broader groups, which are summarized below.

(a) *Hydrological Modeling Approaches*

Hydrological modeling refers to the process of simulating the rainfall–runoff relationship within a catchment by integrating hydrological, meteorological, and land surface data. It provides a structured, process-based means of reproducing the flows and thus its is conveniently used to assess how the flows get altered by climate variations and human interventions.

It is one of the most widely adopted approaches for separating the effects of climate variability and human activities on runoff. The application of this method begins with the identification of a reference period and an interference period, which serve as the foundation of the attribution framework. Change-point detection techniques such as the Pettitt test, Mann–Kendall rank test, sequential Mann–Kendall test, moving t-test, or double-mass curve analysis are commonly used to determine the point in time when runoff characteristics shift significantly, thereby dividing the dataset into two distinct phases. The reference period is assumed to represent natural runoff conditions with minimal human influence, while the interference period captures the altered flow regime under the combined influence of climate and human activities.

Once these periods are established, hydrological models are calibrated and validated using data from the reference period. These models can be broadly classified as lumped models, which treat the catchment as a single homogeneous unit, or distributed models, which explicitly account for spatial heterogeneity in land cover, soils, and topography. Examples of lumped models include SIMHYD (Liang et al. 2013), TOPKAPI (Ciarapica and Todini 2002), DiCaSM (Montenegro and Ragab 2010), DTVGM (Wang et al. 2009), and the Xinanjiang model (Ahirwar et al. 2018). Distributed models include RHESSys

(López-Moreno et al. 2014), SWAT (Li et al. 2021), VIC (Kang et al. 2022), GBHM (Xu et al. 2013), DPBM (Refsgaard et al. 1989), TOPMODEL (Nourani et al. 2011), DHSVM (Wigmosta and Lettenmaier 1999), and PRMS (Legesse et al. 2003), which have been widely applied in diverse regions.

After calibration, the models are applied to the interference period using meteorological inputs only, thereby simulating the runoff that would have occurred if human activities were absent. The difference between the simulated and observed runoff during the interference period quantifies the impact of human activities, whereas the difference between the simulated runoff in the interference period and the observed runoff in the reference period reflects the influence of climate variability. This attribution is mathematically expressed as:

$$Q_{total} = Q_h + Q_c \tag{2.1}$$

$$\Delta Q_h = Q_{post} - Q_{sim} \tag{2.2}$$

$$\Delta Q_c = Q_{sim} - Q_{pre} \tag{2.3}$$

where ΔQ_{total} is the total change in runoff between the reference and interference periods, ΔQ_h is the change in runoff caused by human activities, ΔQ_c is the change in runoff caused by climate variability, Q_{post} is the observed runoff in the interference period, Q_{sim} is the simulated runoff in the interference period, and Q_{pre} is the observed runoff in the reference period.

Hydrological modeling offers several strengths. It provides a process-based and quantitative framework for attribution, is flexible across scales from small catchments to large basins, and has been extensively tested under diverse hydro-climatic conditions. Distributed models, in particular, capture spatial heterogeneity in climate and land surface characteristics. However, the method is also constrained by certain limitations. Results are highly sensitive to input data quality and model calibration, while the extrapolation of point-based meteorological records to basin scales can introduce uncertainty. Moreover, model structures and parameterizations may not fully represent real hydrological processes, and distributed models can be computationally intensive, especially for large basins.

A refinement of this framework is the scenario combination approach, which strengthens attribution by running hydrological models under different combinations of land use and land cover (LULC) and climate inputs (Li et al. 2009). The equations are expressed as:

$$Total\ Change = Sim1 - Sim2 \tag{2.4}$$

$$\%Change\ due\ to\ LULC = \left(\frac{Sim2 - Sim3}{Total\ Change}\right) \times 100 \tag{2.5}$$

$$\%Change\ due\ to\ Climate = \left(\frac{Sim1 - Sim3}{Total\ Change}\right) \times 100 \tag{2.6}$$

Here, Sim1 represents the simulated runoff for the reference period, Sim2 for the interference period, and Sim3 is generated by running the model with reference-period LULC but interference-period meteorological data. The difference between Sim1 and Sim3 reflects the influence of climate, because LULC is held constant at reference conditions while meteorological inputs vary across periods, and dividing this difference by the total change expresses it as a proportion of overall runoff alteration. Similarly, the difference between Sim2 and Sim3 isolates the effect of LULC change, since climate inputs remain fixed while land use shifts between the reference and interference periods. By comparing both proportions to the total change, the scenario combination approach provides a clear and percentage-based attribution of runoff alteration, allowing a sharper separation of climate-driven and human-driven impacts. This method is particularly useful in basins undergoing rapid LULC transitions; however, its effectiveness depends on the availability of accurate LULC and climate datasets, and the increased complexity may amplify uncertainties in data-scarce regions.

Hydrological modeling-supported by the scenario combination refinement-remains one of the most powerful and widely used methods for quantifying the relative roles of climate variability and human activities in altering runoff. Crucially, the identification of reference and interference periods through robust change-point detection serves as the cornerstone of this approach, ensuring that attribution is grounded in the temporal dynamics of the hydrological record.

(b) *Conceptual Approaches*

The conceptual category of methods is dominated by the Budyko framework and its extensions, alongside the Tomer-Schilling approach, both of which offer simple yet powerful ways to partition the influence of climate variability and human activities on runoff.

The Budyko hypothesis (Budyko 1961; Budyko and Miller 1974) provides a physically based representation of the long-term water-energy balance of a catchment. It posits that actual evapotranspiration (E) is jointly constrained by water supply, represented by precipitation (P), and energy availability, represented by potential evapotranspiration (ET_0). In arid regions where ET_0 far exceeds P, evapotranspiration is water-limited and $E \approx P$, whereas in humid regions with abundant rainfall, evapotranspiration is energy-limited and $E \approx ET_0$. The ratio ET_0/P defines the aridity index, while E/P defines the evapotranspiration index (Henning and Flohn 1977). To account for basin heterogeneity, parameters such as ω, n, or w have been introduced, representing soil, vegetation, topography, or climatic seasonality (Milly 1994; Donohue et al. 2007; Yang et al. 2008; Potter et al. 2005).

The Budyko function can therefore be generalized as:

$$\frac{E}{P} = F\left(\frac{ET_0}{P}\right), \ c \tag{2.7}$$

where c denotes catchment properties, with different values of c yielding different forms of Budyko-type curves (Zhou et al. 2015).

Two attribution techniques have developed within this framework. The elasticity method (Schaake 1990) quantifies the sensitivity of runoff to climate drivers through elasticity coefficients, defined as:

$$\varepsilon_x = \lim_{\Delta x/x \to 0} \frac{\Delta Q/Q}{\Delta x/x} = \frac{\partial Q}{\partial x} \times \frac{x}{Q} \tag{2.8}$$

An elasticity of 0.1, for instance, implies that a 10% change in a variable such as precipitation or ET_0 produces a 1% change in runoff. Starting from the water balance:

$$P = E + Q \tag{2.9}$$

we obtain:

$$\frac{dQ_c}{Q} = \varepsilon p \frac{dP}{P} + \varepsilon ET_0 \frac{dET_0}{ET_0} \tag{2.10}$$

Or in discrete form:

$$\frac{\Delta Q_c}{Q} = \varepsilon p \frac{\Delta P}{P} + \varepsilon ET_0 \frac{\Delta ET_0}{ET_0} \tag{2.11}$$

which isolates the climate-driven component (ΔQ_c) of runoff alteration.

The decomposition method (Wang and Hejazi 2011) extends the Budyko hypothesis by distinguishing between movements along the Budyko curve, which represent climate variability, and vertical departures from the curve, which represent human activities. In this view, precipitation and ET_0 drive changes in climatic water balance along the curve, while LULC change, irrigation, or reservoir regulation shift evapotranspiration independently of climate, producing vertical displacements. Human-induced change is expressed as:

$$\Delta Q_h = \left(\frac{E_2}{P_2} - \frac{E_2'}{P_2}\right) P_2 \tag{2.12}$$

where E_2 is observed evapotranspiration in the interference period, E_2' is Budyko-predicted evapotranspiration for the same period, and P_2 is precipitation. The difference between actual and predicted evapotranspiration thus isolates the contribution of human activities.

While the Budyko framework is conceptually elegant, data-light, and highly effective at basin and regional scales, it is less suitable for short-term or scenario-based analyses, and depends heavily on the accuracy of ET_0 estimates (Penman 1948; Teng et al. 2012).

Complementing this, the Tomer–Schilling framework (Tomer and Schilling 2009) offers a diagnostic approach to evaluate ecosystem water-energy partitioning. It introduces two indices-water excess (P_{ex}) and energy excess (E_{ex})-defined as:

$$E_{ex} = \frac{PE - AE}{PE} \tag{2.13}$$

$$P_{ex} = \frac{P - AE}{P} \tag{2.14}$$

where PE is potential evapotranspiration, AE is actual evapotranspiration, and P is precipitation. Plotting P_{ex} against E_{ex} provides a graphical means of diagnosing whether hydrological alterations are driven by climate change (movements along the energy-water tradeoff) or by human activities such as land use and management changes (orthogonal shifts). Although it does not directly compute runoff changes, the framework helps disentangle underlying drivers of hydrological variation. Its main limitation lies in its data requirements-long-term, high-quality records of precipitation, evapotranspiration, and meteorological parameters are essential-and in its simplifying assumption that human activities affect AE independently of climate (Ye et al. 2013).

Budyko and Tomer–Schilling frameworks represent the core of conceptual approaches in runoff attribution. Both begin with the separation of a reference period and an interference period, identified through change-point detection techniques, and then use long-term climatic and hydrological balances to partition observed runoff changes between climate variability and human influence. Importantly, this reliance on reference versus interference periods is not unique to conceptual approaches but is a unifying feature across almost all attribution methods-whether experimental, statistical, modeling-based, or analytical-because only by establishing this temporal boundary can the distinct roles of climate change and human activities be meaningfully disentangled. Their value lies in their parsimony and diagnostic clarity, though their accuracy ultimately depends on the availability and quality of climatic and evapotranspiration data.

(c) *Analytical Approaches*

Analytical methods provide a compact and mathematically rigorous means of attributing runoff changes to climate variability and human activities, primarily through climate elasticity and hydrological sensitivity frameworks. Like other categories of attribution, these approaches rely on identifying a reference period and an interference period, separated by a change point, to isolate natural variability from human-induced effects.

The climate elasticity concept, introduced by Schaake (1990), measures how sensitive streamflow (Q) is to changes in climate drivers, particularly precipitation (P) and potential evapotranspiration (PE). Elasticity is defined as the ratio of the proportional change in streamflow to the proportional change in a climate variable:

$$\varepsilon_P = \frac{\frac{dQ}{Q}}{\frac{dP}{P}} = \frac{dQ}{dP} = \frac{P}{Q} \tag{2.15}$$

Sankarasubramanian et al. (2001) improved this approach by developing a non-parametric estimator of precipitation elasticity, reducing bias and avoiding strong model assumptions:

$$\varepsilon_P = median\left(\frac{Q - \overline{Q}}{P - \overline{P}}\right)\frac{Q}{P} \tag{2.16}$$

While this formulation is less sensitive to model structure, it initially neglected the role of temperature. Fu et al. (2007) therefore proposed an extension incorporating both precipitation and temperature effects:

$$\varepsilon_P = \left(\frac{Q_{P,\delta T} - \overline{Q}}{P_{P,\delta T} - \overline{P}}\right)\frac{\overline{P}}{\overline{Q}} \tag{2.17}$$

where $\delta T = T - \overline{T}$ represents temperature deviation from its mean, capturing nonlinear precipitation–temperature–runoff relationships.

For small-sample basins, Zheng et al. (2009) addressed statistical limitations by introducing a least-squares estimator:

$$\varepsilon_P = \frac{\overline{X}}{\overline{Q}}\frac{\sum(X_i - \overline{X})(Q_i - \overline{Q})}{\sum(X_i - \overline{X})^2} = \rho_{X,Q}\frac{C_Q}{C_X} \tag{2.18}$$

where X is a climate variable, $\rho(X, Q)$ is the correlation between X and Q, and C_X, C_Q are coefficients of variation.

More comprehensive extensions were introduced by Yang and Yang (2011), who incorporated precipitation (P), temperature (T), net radiation (Rn), and wind speed (U_2):

$$\frac{dR}{R} = \varepsilon_a\frac{dP}{P} + \varepsilon_b\frac{dT}{T} + \varepsilon_c\frac{dR_n}{R_n} + \varepsilon_d\frac{dU_2}{U_2} \tag{2.19}$$

Liu et al. (2013) further reformulated elasticity in terms of the aridity index ($\varphi = PE/P$), deriving a sensitivity coefficient:

$$S_\varphi = \lim_{\frac{\Delta\varphi}{\varphi}\to 0}\left(\frac{\frac{\Delta Q}{Q}}{\frac{\Delta\varphi}{\varphi}}\right) = \frac{\partial Q}{\partial\varphi}\frac{\varphi}{Q} \tag{2.20}$$

which leads to expressions for the contribution (C_φ) and rate of contribution (P_φ) of aridity change:

$$C_\varphi = S_\varphi \times \frac{\varphi_2 - \varphi_1}{\varphi_1} \tag{2.21}$$

$$P_\varphi = \frac{C_\varphi}{\frac{Q_2 - Q_1}{Q_1}} = \frac{S_\varphi \times \frac{\varphi_2 - \varphi_1}{\varphi_1}}{\frac{Q_2 - Q_1}{Q_1}} \tag{2.22}$$

Thus, climate elasticity methods allow the proportional influence of precipitation, temperature, and aridity on runoff to be quantified, though their performance depends on long-term, high-quality climate–streamflow datasets.

The hydrological sensitivity method is closely related to climate elasticity, but it explicitly links changes in mean streamflow to precipitation and evapotranspiration perturbations. Milly and Dunne (2002) expressed climate-driven changes in annual mean streamflow (ΔQc) as:

$$\Delta Q^C = \beta \Delta P + \gamma \Delta PE \tag{2.23}$$

where β and γ represent streamflow sensitivity to precipitation and potential evapotranspiration, respectively. The total change in runoff between reference and interference periods is decomposed as:

$$\Delta Q^{total} = \overline{Q_2} - \overline{Q_1} = \Delta Q^C + \Delta Q^H \tag{2.24}$$

with ΔQ^H representing human-induced changes. This formulation assumes pre-change point variability is climate-driven, while post-change alterations combine climate and human effects.

Building on this, Zhang et al. (2001) introduced a two-parameter model linking actual evapotranspiration (AE) to rainfall (P), potential evapotranspiration (PE), and plant-available water capacity (w):

$$\frac{AE}{P} = \frac{1 + W\frac{PE}{P}}{1 + W\frac{PE}{P} + \left(\frac{PE}{P}\right)^{-1}} \tag{2.25}$$

with sensitivity coefficients defined as (Li et al. 2007):

$$\beta = \frac{1 + 2\frac{PE}{P} + 3W\frac{PE}{P}}{\left(1 + \frac{PE}{P} + W\left(\frac{PE}{P}\right)^2\right)^2} \tag{2.26}$$

$$\gamma = -\frac{1 + 2W\frac{PE}{P}}{\left(1 + \frac{PE}{P} + W\left(\frac{PE}{P}\right)^2\right)^2} \tag{2.27}$$

The relative contributions of climate variability (Q_C) and human activities (Q_H) to total runoff change are then expressed as:

$$Q_C = \frac{\Delta Q^C}{\Delta Q^{total}} \frac{1 + 2W\frac{PE}{P}}{\left(1 + \frac{PE}{P} + W\left(\frac{PE}{P}\right)^2\right)^2} \tag{2.28}$$

$$Q_H = \frac{\Delta Q^H}{\Delta Q^{total}} \frac{1 + 2W\frac{PE}{P}}{\left(1 + \frac{PE}{P} + W\left(\frac{PE}{P}\right)^2\right)^2} \tag{2.29}$$

Analytical approaches are advantageous because they provide explicit mathematical relationships between climate drivers and streamflow response, require relatively modest data inputs compared to complex hydrological models, and yield interpretable sensitivity metrics. However, their accuracy depends heavily on long-term hydrometeorological records and their assumptions of linearity may oversimplify basin-scale processes. Still, when coupled with change-point detection to separate reference and interference periods, they offer a powerful, transparent framework for disentangling climate- versus human-induced runoff alterations.

(d) *Experimental Approaches*

Experimental approaches provide some of the earliest and most direct methods for quantifying the influence of human activities on runoff, relying primarily on long-term hydrological field observations. The classic example is the paired catchment method, in which two adjacent catchments with similar physical characteristics-such as size, soils, topography, and vegetation-are monitored simultaneously. One catchment serves as the control, while the other undergoes a treatment such as deforestation, afforestation, or land-use change. During a calibration period, a regression relationship between the streamflows of the two catchments is established under undisturbed conditions. Following the intervention in the treatment catchment, any systematic deviation from this established relationship is attributed to human activities. This approach has been particularly effective in small experimental catchments (generally <1 km^2), where hydrological responses to vegetation change can be closely monitored (Zhao et al. 2009). However, its wider applicability is constrained by several limitations: it becomes unreliable in larger basins (100–2500 km^2) due to heterogeneity in catchment properties, requires long-term observation periods, and is often hindered by the practical difficulty of finding truly comparable catchments under identical climatic conditions (Lørup et al. 1998; Zhao et al. 2010; Li et al. 2009).

Where no suitable control catchment exists, the time-trend method offers an alternative. Here, a regression relationship is first established between precipitation (P_1) and streamflow (Q_1) during a calibration period:

$$Q_1 = f(P_1) \tag{2.30}$$

The same regression function is then applied to precipitation data (P_2) from a subsequent testing period to estimate the natural streamflow (Q_2) that would have occurred in the absence of human disturbance:

$$Q_2 = f(P_2) \tag{2.31}$$

The difference between the observed mean streamflow ($\overline{Q}$) during the testing period and the predicted natural streamflow ($\overline{Q}_2$) is attributed to human activities:

$$\Delta Q^{veg} = \overline{Q} - \overline{Q_2} \tag{2.32}$$

The underlying principle is straightforward: if climate is the sole driver of hydrological change, observed streamflow should closely match the predicted value. Any systematic departure between the two therefore reflects human-induced alterations such as deforestation, afforestation, or agricultural land management, which modify evapotranspiration, infiltration, and runoff processes. While the precipitation–streamflow relationship may be nonlinear (sometimes modeled with hyperbolic tangent functions), most applications adopt linear regression due to its simplicity and interpretability (Grayson et al. 1996). A key drawback of the time-trend method, however, is that it typically accounts only for climatic variables and neglects geomorphological characteristics that also influence runoff generation.

Despite their limitations, experimental approaches remain valuable for their simplicity and their grounding in direct hydrological observations. Like other attribution frameworks, they depend on the careful identification of a reference period and an interference period to separate natural variability from human influence. In the paired catchment method, this distinction arises from the calibration phase before land-use intervention versus the treatment phase after intervention, while in the time-trend method, calibration and testing phases serve the same role. Unlike statistical or modeling approaches, however, these methods do not typically apply formal change-point detection tests; rather, the "change point" is predetermined by the known timing of land-use or management changes.

(e) *Empirical Approach*

Empirical approaches are primarily based on statistical methods which is widely applied to investigate the combined effects of climate variability and human activities on runoff. These methods rely on long-term hydrological and meteorological datasets to establish statistical relationships between streamflow and climatic drivers. A relationship is first constructed during a reference period, which reflects natural hydrological conditions with minimal human disturbance. This calibrated model is then applied to the interference period, using meteorological data alone to simulate the "natural" runoff that would have occurred in the absence of human influence. By comparing the simulated runoff with the observed runoff during the interference period, the respective contributions of climate variability and human activities can be disentangled.

Mathematically, this framework can be expressed as:

$$Q_{pre} = \int (C_{pre}) \tag{2.33}$$

$$Q_{sim} = \int (C_{post}) \tag{2.34}$$

where Q_{pre} represents the natural runoff simulated using climate variables (C_{pre}) from the reference period, and Q_{sim} represents the simulated runoff for the interference period using climate variables (C_{post}). The difference between Q_{sim} and the observed runoff in the interference period is attributed to human activities, while deviations between Q_{pre} and Q_{sim} capture the influence of climate variability.

A wide range of statistical techniques have been employed within this framework. Early studies used simple statistics (Costa et al. 2003) and linear regression (Ahn and Merwade 2014; Saifullah et al. 2016), while more sophisticated applications involved multiple regression (Jiang et al. 2011b), stepwise multiple regression (Xu 2011; Du and Shi 2012), and fuzzy linear regression (Sahin and Hall 1996). Other techniques include the flow duration curve method (Wu et al., 2011), the double-mass curve (Jiang et al. 2011a; Xue et al. 2017), its modified version (Wei and Zhang 2010), and baseflow recession analysis (Wang and Cai 2010). Each of these approaches applies statistical models to assess the extent of runoff alteration in response to climatic shifts and land-use changes.

The strength of empirical statistics lies in their simplicity and modest data requirements. Only runoff observations and basic meteorological variables are necessary, making the method practical and accessible, especially in data-limited contexts. However, their limitations are equally notable. Because they are based purely on correlations rather than hydrological processes, they lack a physical mechanism to explain runoff changes. As such, while they can effectively capture the impacts of climate variability, they provide only indirect evidence of human contributions. Moreover, their accuracy is strongly dependent on the quality and length of the input datasets.

Empirical statistical methods provide a pragmatic yet process-light approach for runoff attribution. By leveraging the distinction between reference and interference periods, they can yield valuable insights into the respective roles of climate and human factors. Nevertheless, their results are most robust when used in combination with conceptual or process-based models that can validate and reinforce their statistical findings.

(f) *Machine Learning*

The application of machine learning (ML) in hydrological studies has expanded rapidly in recent years with the growth of big data technologies (Ardabili et al. 2020). Unlike traditional process-based or statistical models, ML does not require explicit mathematical representations of hydrological processes but instead extracts patterns and relationships directly from large datasets (Shen 2018; Chemura et al. 2020). This makes it particularly valuable in modeling runoff, where nonlinear interactions between climatic drivers and catchment characteristics are often complex and poorly understood. In addition, ML can overcome some limitations of traditional models, such as computational cost, parameter uncertainty, and sensitivity to data gaps (Zeng et al. 2020).

Early studies demonstrated the promise of artificial neural networks (ANN) for runoff prediction (Karunanithi et al. 1994; Lorrai and Sechi 1995). Since then, a range of ML approaches has been successfully applied to distinguish the relative impacts of climate variability and human activities on streamflow. Examples include ANN (Bai et al. 2016; Tongal and Booij 2018), deep belief networks (DBN) (Bai et al. 2016), support vector regression (SVR) (Tongal and Booij 2018), and random forest (RF) (Tongal and Booij 2018). Kratzert et al. (2018) advanced this framework by training ML models on reference-period data to simulate natural runoff under climatic conditions. These trained models were then applied to interference periods affected by human disturbance. Deviations between simulated and observed runoff in the interference period were attributed to human activities, while differences between simulated and observed runoff in the reference period were linked to climate variability. This approach highlights the ability of ML to disentangle climate- and human-driven contributions in catchment hydrology.

Despite these advantages, ML applications in hydrology face certain challenges. Reliable model development requires sufficiently long and high-quality time series, with Kratzert et al. (2018) recommending at least 15 years of daily data to avoid overfitting. In addition, ML approaches may lack physical interpretability, as they capture statistical associations rather than mechanistic processes (Zeng et al. 2020). Consequently, while machine learning provides a powerful complement to traditional hydrological methods, its use must be carefully constrained by data availability and ideally combined with process-based insights to ensure robustness and interpretability.

However, it is worth noting that the selection of an appropriate model to assess the impacts of climate change and human activities on runoff alteration depends on the study area, available datasets, research goals, and model assumptions. Different models may have distinct advantages and disadvantages in simulating the runoff responses (Table 2.3). Thus, it is essential to assess the performance and suitability of various models in a specific study area before choosing the suitable model for the research objectives.

The evolution of methods for assessing runoff alteration reflects a clear shift from empirical and observational approaches (e.g., paired catchment studies and regression-based analysis) to advanced, simulation-driven frameworks. Early research primarily relied on statistical trend detection (e.g., Mann–Kendall, Pettitt) and simplified water balance models to examine runoff responses to climatic or anthropogenic drivers. With advancements in geospatial technology and process-based hydrological modeling, particularly the adoption of Soil and Water Assessment ToolSWAT), studies now employ more comprehensive, scenario-based analyses of climate and land-use impacts. The

Table 2.3 Strengths and weaknesses of the models used to separate the impact of climate change and human activities on runoff alterations

Sl. No.	Name of the model	Strengths	Weaknesses
1.	Hydrological modeling approaches (including scenario combination methods)	a. It can be applied across a wide range of spatial and temporal scales, from small to large basins and from hourly to annual simulations b. It can be calibrated and validated using observed data, improving reliability c. It can be integrated with climate models and land use models to assess combined impacts d. It is cost-effective compared to long-term experimental studies, avoiding the need for expensive and time-consuming fieldwork e. It can incorporate multiple climate and land use scenarios, thereby separating climate- and human-induced contributions while exploring future uncertainties through scenario testing	a. It requires high-resolution hydro-meteorological data, and interpolation from point stations often introduces uncertainty b. Its accuracy depends on parameterization and calibration of equations and algorithms c. It is complex to develop, calibrate, and validate d. It may not be suitable for highly complex or significantly altered catchments, and results can be sensitive to scenario assumptions

(continued)

Table 2.3 (continued)

Sl. No.	Name of the model		Strengths	Weaknesses
2.	Conceptual approaches	Budyko hypothesis (decomposition and sensitivity method)	a. It is applicable in data-scarce regions b. It has a sound physical basis c. It is a relatively simple approach that requires only a few inputs variables d. It is used across a range of hydrological conditions and geographical locations e. It can be applied to different spatial and temporal scales, from small catchments to large river basins	a. There is no system in place to verify the accuracy of the ET_0 calculations b. It is although globally applicable, accuracy may be reduced in humid or snow-dominated catchments c. It relies on the assumption of steady-state conditions and vegetation is the main control of evapotranspiration. These assumptions may not hold in all catchments, which can lead to inaccurate conclusions
		Tomer Schilling framework	a. It can help to better capture the complexity of factors influencing streamflow b. It can help to better identify long-term trends c. It can be integrated with existing hydrologic and climate model	a. It does not provide direct insights into runoff variation b. It requires long-term datasets and high-resolution data on a range of hydrological variables, including rainfall, evapotranspiration, and soil moisture c. It is designed for use in urban and suburban catchments, which may not be representative of other types of catchments

(continued)

Table 2.3 (continued)

Sl. No.	Name of the model		Strengths	Weaknesses
3.	Analytical approaches	Climate elasticity model	a. It provides a direct and quantitative measure of streamflow sensitivity to climate drivers b. It can help to identify the impacts of long-term climate change c. It can be applied at various scales d. It can be integrated with existing hydrologic and climate models	a. It does not consider precipitation distribution, frequency, or dynamic changes in underlying surface conditions b. It may not apply to all types of catchments or all types of human activities c. Catchment behavior may vary spatially, which can make it difficult to generalize the results to other areas d. It requires detailed data on a range of hydrological variables, including rainfall, evapotranspiration, and soil moisture
		Hydrological sensitivity method	a. It can be applied to different spatial and temporal scales b. It can predict accurate and reliable future changes in runoff	a. It requires long-term datasets on a range of hydrological variables, including rainfall, evapotranspiration, and soil moisture b. There may be other factors that are contributing to the changes that are not accounted for c. Catchment behavior may vary spatially, which can make it difficult to generalize the results to other areas

(continued)

Table 2.3 (continued)

Sl. No.	Name of the model		Strengths	Weaknesses
4.	Experimental methods based on hydrological field data	Paired catchment approach	a. It enables direct comparison between two catchments b. It can be a cost-effective method c. It is repeatable and can be used over time to monitor changes in runoff d. The impact of confounding variables (geology, topography, and climate) can be minimized e. It can provide robust evidence to support decision-making processes	a. It is only applicable in a small basin and needs high-quality data b. It is a time-consuming approach c. It is difficult to extrapolate to other basins d. It can be challenging to find two catchments that are truly comparable e. It requires a relatively large amount of data to be collected from each catchment f. Extrapolation to other basins or regions is limited, as results are highly site-specific, even though local causal inference is strong

(continued)

Table 2.3 (continued)

Sl. No.	Name of the model		Strengths	Weaknesses
		Time trend method	a. It is useful for assessing the impact of vegetation change in large basins b. It is a straightforward statistical technique and easy to implement c. It is a data-driven approach d. It can be used to analyze long-term trends e. It can be applied to different time scales (annual to multi-decadal) f. It can be combined with other statistical techniques (regression and time series modeling)	a. There may be other factors that are contributing to the changes that are not accounted for b. It may not be able to detect significant changes in the runoff if the data is too noisy or if the changes are too small to be detected with the available data c. It requires high-quality data to produce accurate results d. It can establish correlations between variables, but it cannot establish causation e. It can be difficult to interpret, reproduce, or replicate results consistently

(continued)

Table 2.3 (continued)

Sl. No.	Name of the model	Strengths	Weaknesses
5.	Empirical statistics	a. It is very simple to implement b. It makes the analysis more objective and less prone to bias compared to subjective methods c. It can provide precise estimates of the magnitude and direction of changes in runoff d. It can be applied to large datasets	a. It lacks a physical mechanism b. It needs high-quality hydro-meteorological data c. It can establish correlations between variables, but it cannot establish causation d. It can be affected by confounding factors that may be difficult to control e. It can be difficult to interpret and replicate
6.	Machine learning	a. It can provide a high level of accuracy b. It has low uncertainty c. It can be applied to large datasets d. It is robust to outliers and noise in the data e. It can be trained on one dataset and then applied to new data f. It can be computationally efficient	a. It requires at least 15 years of daily data, otherwise can lead to inaccurate predictions b. It requires careful consideration of the datasets c. It often lacks physical interpretability, making it challenging to understand or verify the underlying processes behind predictions d. It requires substantial domain expertise for reliable application and interpretation

Source Tabulated by authors

integration of trend analysis with SWAT simulations exemplifies a methodological convergence that not only quantifies historical changes but also enables predictive assessments under future conditions. This evolution marks a transition from descriptive to diagnostic and prescriptive hydrologic analysis, providing the methodological foundation for this study. Importantly, such integrated approaches also offer the potential to explore upstream–downstream linkages, thereby connecting basin-scale hydrologic alterations to their cascading impacts on deltaic systems, an area that remains underexplored in current literature.

2.2.3 Review of Research Outcomes

Close examination of all selected literature reveals that runoff response to changing climate and human activities is largely controlled by factors such as the spatial scale of the basin, types of human activities, climate types, geophysical characteristics of the basin, etc., each of which is reported below in brief (Appendices A.1 and A.2).

(a) *Spatial Scale and Runoff Response*

Generally, larger basins tend to be more resilient to climate variability than smaller ones (Palmer et al. 2008). On the other hand, changes in land use and land cover over smaller basins may have a more significant impact on the hydrology than in larger basins (Dos Santos et al. 2018). To address this scale-dependent variability, the present study classifies river basins into nine categories based on their spatial extent, following the framework proposed by the All India Soil and Land Use Survey (1990) and Singh (1994), as illustrated in Fig. 2.1.

In their findings in China, Li et al. (2009), Xu et al. (2014), Yang et al. (2014), Zhang et al. (2014a, b), Wang et al. (2017, 2020) showed that the climate variability caused reduction in runoff by about 19–95.8%, while human activities accounted for change in runoff by 40–81%. These wide range of variation in relative contribution of climate change and human activities is due to variation in size of the basin under investigation.

Similar kind of studies were carried out by Beguería et al. (2003), Wang et al. (2009), Bao et al. (2012), Zhang et al. (2017), Pirnia et al. (2019), Andaryani et al. (2021) and many others in different parts of the globe at the sub-catchment level. They revealed that climate variability was responsible for reducing the runoff by 70%, 29.9%, 42%, 39.1%, 29.86–40.19%, and 83.6%, respectively. On the other hand, human activities accounted for a reduction in runoff by 30%, 70.1%, 59%, 63.1–64.8%, 59.81–70.14%, and 16.4%, respectively. As per their reporting, all these differences are due to variation in spatial extent of catchment under study.

Appendices A.1 and A.2 show the list of selected literature from different parts of the globe which have reported the variation in catchment response caused by climate change and human activities across the basins of different spatial extents.

(b) *Types of Human Activities and Runoff Response*

The pattern of runoff responses was also influenced by the degree of human activity that alters the landscape by modifying the LULC under changing climate conditions. Studies reviewed in this research have shown that the LULC of basins can be agricultural, forest, grassland, mixed, and urban (Appendices A.1 and A.2).

Several research studies carried out by Chen et al. (2005), Wang et al. (2009, 2020), Bao et al. (2012), and Liang et al. (2015) over agriculture dominated basins in various locations have revealed that climate variability was responsible for 60–80, 29.9, 26, 38, and 60% reduction in runoff in the respective area while reduction in runoff by about 20, 70.1, 74, 62, and 40% were attributed to human activities. So, climate change and human activities have led to offsetting responses in agriculture-dominated basins.

Over forested basin, as reported by Beguería et al. (2003), Liu et al. (2010), Li et al. (2013), and Zhang et al. (2020), the climate variability contributed to the reduction in the runoff by 70%, 50%, 45%, and 83% respectively whereas it was respectively 30, 50, 55, and 17% due to human activities (Appendices A.1 and A.2). On the other hand, the contribution rate of climate variability (ranging from 83.58 to 90.5%) and human activities (ranging from 9.5 to 16.42%) to the reduction of runoff in grassland basins were found to be almost consistent in different areas of the world as reported by Chen et al. (2013), Shang et al. (2019), Andaryani et al. (2019), and Yan et al. (2020). For urban catchment, as reported by Aboelnour et al. (2019) for Little Eagle Creek in the USA, climate change has a higher contribution rate than human activities. According to the studies conducted by Fan et al. (2010), Bao et al. (2012), Xu et al. (2014), and Tang et al. (2021), mixed basins have shown different responses, but in all cases, human activities had a higher impact than climate variability in reducing runoff (Appendices A.1 and A.2).

In many research outcomes, it is also observed that there is an increase in runoff due to climate change, whereas human activities have caused runoff reduction, although their relative contribution differs across the basin of different LULCs. The notable researchers in this context are Ma et al. (2009), and Shi et al. (2013) (for agricultural watershed), Zhou et al. (2020) (for forested watershed), Zhang et al. (2018) (for grassland basin), Gibson et al. (2006), and Khoi et al. (2019) (for mixed basin) (Appendices A.1 and A.2).

(c) *Climatic Types and Runoff Response*

The literature survey reveals that climate variability and change in human activities have positive as well as negative impacts on runoff, and there is wide variation in the relative contribution of both factors across the climatic zones.

In the tropical region, the contribution of human activities in the reduction of runoff ranges from 54.19 to 68.39%, whereas the percentage share caused by climate change is 31.61–45.81%, as reported by Wang et al. (2012, 2021), Zhang et al. (2020), and many others (Appendices A.1 and A.2). Exact opposite results are found for the sub-tropical region, as reported by Chen et al. (2005), Zhang et al. (2017), and Zhou et al. (2018), and the same scenario is also found for the region having transitional climate from semi-humid to semi-arid region as reported by Li et al. (2010), Bao et al. (2012), and many others.

In many cases, the scenario over purely semi-arid regions is found to be consistent with what is observed in the tropical region, i.e., in the semi-arid region, the percentage share of climate change impact on reducing runoff was relatively small as compared to that caused by the change in human activities as reported by Li et al. (2007), Fan et al. (2010), and Seyoum et al. (2015) (Appendices A.1 and A.2). At the same time, there are many examples of just the opposite scenario for semi-arid and arid regions, e.g., the Rwizi catchment in Uganda (Onyutha et al. 2021), Kaidu River basin in China (Chen et al. 2013), wherein climate variability was the main contributor to the reduction in runoff.

Many studies reported the positive impact of climate variability and human activities on runoff, i.e., both factors cause an increase in runoff (Tomer and Schilling 2009; Juckem et al. 2008; Heo et al. 2015; Napoli et al. 2017; Yang et al. 2019; Nkhoma et al. 2021). Many authors, on the other hand, observed in many places that there is an increase in runoff due to climate variability and a decrease due to human activities (Ma et al. 2009; Ye et al. 2013; Yang et al. 2017). Just opposite results are also reported from many places by Peña-Arancibia et al. (2012) and Zhou et al. (2020). All these are experienced by the region irrespective of climate type.

(d) *Geophysical Conditions and Runoff Response*

Geophysical condition is hardly found to have any control over runoff response to changing climate and human activities. Hence, regions with similar geophysical conditions, i.e., regions having similar relief and similar physiography, show different runoff responses (Wang et al. 2010, 2017; Seyoum et al. 2015; Zhang et al. 2018, 2021; Khoi et al. 2019) (Appendices A.1 and A.2). The response, here, depends upon the magnitude and extent of change in climate and human activities, which may occur irrespective of surface characteristics in basin scale hydrologic system.

Collectively, these studies confirm that runoff alteration is driven by interconnected climatic, anthropogenic, and physiographic factors, with their relative influence varying by scale and context. Advances in process-based models (e.g., SWAT), coupled with statistical trend and change-point analyses, have enhanced the ability to quantify and disentangle these drivers.

However, despite extensive progress, most research remains confined to basin-scale hydrology, seldom tracing how upstream runoff alterations cascade downstream into deltaic systems. This gap is particularly significant in monsoon-fed tropical basins, where upstream flow and sediment changes have profound implications for delta sustainability and coastal resilience.

2.2.4 Identifying the Research Gap

There has been a lot of methodological advancement in the field of runoff alteration studies under changing climatic conditions and human activities, starting with the paired catchment approach, followed by empirical statistical methods, and more recently by physically based hydrological models integrated with machine learning and artificial intelligence. However, the fundamental field of inquiry has remained the same, i.e., understanding the functional relationship among runoff and other hydrological variables within the broader framework of the hydrological system.

Till the present, there is no satisfactory answer to the question—Why is the hydrologic response of one basin not always found to be consistent with that of another one with similar geophysical conditions? There is also no consensus regarding the relative contribution of climate variability and human activity to cause a change in system response. Moreover, the number of studies is not sufficient to state how the difference in the spatial scale of the hydrologic system affects the relationship among hydrologic variables (Appendices A.1 and A.2).

So, the knowledge gap exists, and to fulfil this gap, more investigation is required on this specific aspect under different geoeconomic environments to understand fully how the runoff generation mechanism is getting changed due to the combined effect of climate change and change in the degree of human interference. The present research is aligned with the same field of enquiry with the intention to find out the regulative principles which the hydrologic system dynamic follows. The focus is on finding the basic mechanism of runoff generation, which is getting changed through the complex interactions between climate change and human activity in the South Koel River basin, which has a contrasting physiographic and cultural landscape under a monsoonal climate.

Additionally, while global research on climate-land use-hydrology interactions is extensive, few studies explicitly examine how upstream hydrologic alterations propagate downstream to impact deltaic systems. By situating the South Koel River Basin within the context of the Brahmani–Baitarani delta system, this study directly addresses this issue, linking basin-scale hydrologic transformations to downstream deltaic resilience-a dimension largely overlooked in prior research which was discussed in next section.

2.3 Review of Upstream–Downstream Hydrologic Connectivity and Deltaic Implications

This section reviews the current state of research on how upstream hydrologic alterations, driven by climate variability and human activities, influence downstream deltaic systems. It emphasizes sediment transport disruption, changes in water flux, and associated ecological consequences, drawing upon global and region-specific case studies.

2.3.1 Sediment and Flow Connectivity Disruption

River basins worldwide are experiencing significant disruptions in sediment and freshwater delivery to deltas due to upstream hydraulic modifications. Reservoirs trap sediments, leading to coastal erosion, delta subsidence, and geomorphic instability. For example, sediment delivery has declined by 30–60% in deltas like the Mississippi, Yangtze, Mekong, and Ganges–Brahmaputra (Darby et al. 2015; Schmitt et al. 2019; Zhang et al. 2022).

2.3.2 Ecological and Geomorphic Impacts of Upstream Modifications

Flow regulation and sediment entrapment reduce floodplain inundation, impair nutrient transport, and threaten wetland fertility and biodiversity. Altered hydrologic pulses fragment aquatic habitats, degrade mangroves, and accelerate salinization in estuarine zones (Xu 2025; Gain et al. 2022).

2.3.3 Internal Delta Connectivity and Hydromorphological Evolution

Within deltaic systems, hydrological and structural connectivity among distributary channels, floodplains, and tidal inlets governs sediment dynamics and nutrient fluxes. Anthropogenic alterations such as levees and embankments sever these linkages, reducing overbank flooding, altering channel morphology, and compromising ecosystem resilience (Hiatt and Passalacqua 2015; Piliouras and Kim 2019).

2.3.4 Modeling Tools for Understanding Connectivity

Advances in modeling frameworks like graph theory and HydroNets offer robust tools to trace sediment pathways, identify vulnerable nodes, and simulate system responses to climate and human pressures. These tools help design adaptive strategies that maintain basin-scale connectivity (Tejedor et al. 2014; Moshe et al. 2020).

2.3.5 Integrated Basin-Delta Management Approaches

Successful resilience planning involves coordinated upstream–downstream management. Examples from the Mekong and Mississippi show how sediment diversions and adaptive dam operations can restore delta-building processes. Strategic interventions must reestablish hydrological continuity while balancing ecological and developmental needs (Darby et al. 2015; Schmitt et al. 2021).

2.3.6 Regional and Global Case Studies

Several major deltas around the world are facing acute sustainability challenges as a result of upstream hydrologic modifications and climate-induced stressors. The Ganga–Brahmaputra–Meghna (GBM) delta, one of the world's largest and most densely populated deltaic systems, is particularly vulnerable to sediment starvation caused by upstream dam construction, water diversions, and projected changes in monsoonal rainfall patterns. These alterations have significantly weakened the natural land-building processes, increasing the risk of coastal erosion, subsidence, and salinity intrusion (Darby et al. 2015). Similar trajectories are observed in other Asian deltas such as Krishna–Godavari, and Mekong deltas.

The Indus Delta has experienced widespread ecosystem collapse and reduced agricultural productivity, largely driven by the upstream diversion of river flows and drastic reductions in sediment supply (Awais et al. 2022). Likewise, the Yellow River Delta in China has undergone substantial morphological degradation and wetland loss, linked to extensive damming and intensive water extraction in the upstream basin (Xu 2025). These consistent patterns emphasize the need for sediment-sensitive and integrated basin governance.

2.3.7 Mountainous Monsoon-Fed Systems and Downstream Vulnerability

In mountainous monsoon-fed regions such as the Himalayas, upstream–downstream hydrologic connectivity is shaped by a complex interplay of land-use change, intense erosion, and highly variable monsoonal precipitation. These dynamics have a direct influence on downstream sediment delivery, water availability, and the timing and intensity of flood regimes which require scale-sensitive management that links headwaters to deltas through Integrated Land and Water Resources Management (ILWRM) approaches (Nepal et al. 2014).

2.3.8 Water Quality and Ecosystem Health

Upstream hydrologic modifications, particularly the construction of large reservoirs, have cascading effects on water quality and ecological health in downstream environments. In tropical regions, reservoirs frequently experience thermal stratification, leading to the release of cooler, oxygen-depleted bottom waters during dam discharges. These altered thermal and chemical regimes can significantly degrade downstream aquatic habitats, affecting species composition, ecosystem productivity, and food web dynamics (Winton et al. 2019).

In addition to thermal and chemical alterations, changes in nutrient transport and flow variability disrupt estuarine dynamics, impacting primary production and nutrient cycling. Moreover, the construction of dams and barrages obstructs natural fish migration routes, particularly for anadromous and estuarine species, leading to habitat fragmentation and declines in biodiversity (Wohl et al. 2021). The combined impact of reduced connectivity, degraded water quality, and ecological isolation poses a significant threat to the long-term health and resilience of river-delta ecosystems.

2.3.9 Connectivity Restoration and Dam Removal

Dam removal has emerged as a viable strategy for restoring hydrological connectivity and rehabilitating riverine and deltaic ecosystems. One of the most well-documented cases is the removal of the Elwha and Glines Canyon Dams on the Elwha River in Washington State, USA. This intervention released ~10.5 million tonnes of stored sediment, which contributed to substantial bed aggradation, channel reformation, and the re-establishment of natural sediment and water flow processes (U.S. Fish and Wildlife Service 2011).

2.3.9.1 Regional Relevance: South Koel and the Brahmani–Baitarani Delta

Despite global advances, monsoon-fed basins in eastern India remain underexplored. The South Koel River's influence on the Brahmani–Baitarani delta's hydrologic, sedimentary, and ecological processes is poorly understood. This study (see Chap. 7) addresses that gap through an integrated upstream–downstream framework, contributing critical insights for adaptive basin-scale management in South Asia.

2.4 Concluding Remarks

This chapter systematically reviewed global and regional research on runoff alteration driven by climate variability and human activities. The review highlighted diverse methodological approaches, ranging from empirical analyses and elasticity-based models to physically-based hydrological modeling, each with its strengths and limitations. It also underscored how runoff responses vary significantly with spatial scale, types of human intervention, climatic regimes, and geophysical conditions.

Despite extensive global research, clear gaps remain particularly in tropical monsoon-fed basins like the South Koel where integrated assessments of climate-LULC interactions are scarce. Moreover, limited attention has been given to linking basin-scale hydrological changes with their downstream impacts on deltaic systems. These gaps underscore the need for a region-specific, process-based investigation that combines statistical analysis, modeling, and spatial techniques to disentangle the relative roles of climate variability and human activities in runoff alteration.

By addressing these gaps, the subsequent chapters focus on empirically examining the South Koel River Basin to provide localized insights and contribute to a broader understanding of upstream–downstream hydrologic connectivity.

Study Area and Research Methodology 3

3.1 Prelude

Understanding the study area and methodological framework is fundamental to any hydrological inquiry, as research outcomes hinge on both geographical context and analytical rigor. The South Koel River Basin, embedded within the Chotanagpur Plateau, represents a dynamic hydro-geomorphic environment characterized by diverse geological formations, climatic variability, and rapidly changing land use and land cover (LULC). This complexity, coupled with its socio-economic importance for agriculture, forestry, and human settlements, makes the basin a critical site for scientific inquiry.

This chapter outlines the basin's physical, climatic, and socio-economic attributes and presents the methodological design, including data sources, pre-processing techniques, hydrological modeling, and statistical analyses. By integrating GIS-based spatial analysis and statistical techniques with simulation modeling, the framework is intended to capture both the spatial heterogeneity and temporal dynamics of runoff alteration, with the overall objective of laying a robust foundation for subsequent analyses, ensuring transparency, reproducibility, and scientific rigor, while establishing a coherent link between earlier conceptual discussions and the empirical assessments that follow.

S. Mondal and R. K. Paul, *Climate Variability and Human Interference*, Deltas of the World, https://doi.org/10.1007/978-3-032-13257-4_3

3.2 The Study Area

3.2.1 Location and Extent

The South Koel River originates near Nagri village in the Ranchi District of Jharkhand at an elevation of about 630 m and joins the river Sankh, and then the combined flow goes by the name Brahmani River. North Karo and South Karo are two important tributaries of the river South Koel. The present study was carried out on the catchment above Jarikela (approx. 22°19′N and 85°5′E), a village located about 50 km downstream of the South Karo–South Koel confluence (CWC 2014). It has an area of 10,356.3 km^2 and is roughly elongated in shape. The basin extends from 21°54′N to 23°37′N and 84°24′E to 85°44′E that encompasses seven districts of Jharkhand (Latehar, Lohardaga, Ranchi, Gumla, Khunti, Simdega, and Paschim Singhbhum) and two districts of Odisha (Kendujhar and Sundargarh) (Fig. 3.1). Most of the basin is in Jharkhand, with only small areas extending into Odisha (Fig. 3.2).

The basin has a maximum length of about 197 km and a width of ~90 km. The trunk stream of the basin, up to Jarikela, spans around 292.90 km, while the length of the North Karo River and South Karo River are 131.94 km and 115.85 km, respectively. The drainage primarily follows a dendritic pattern.

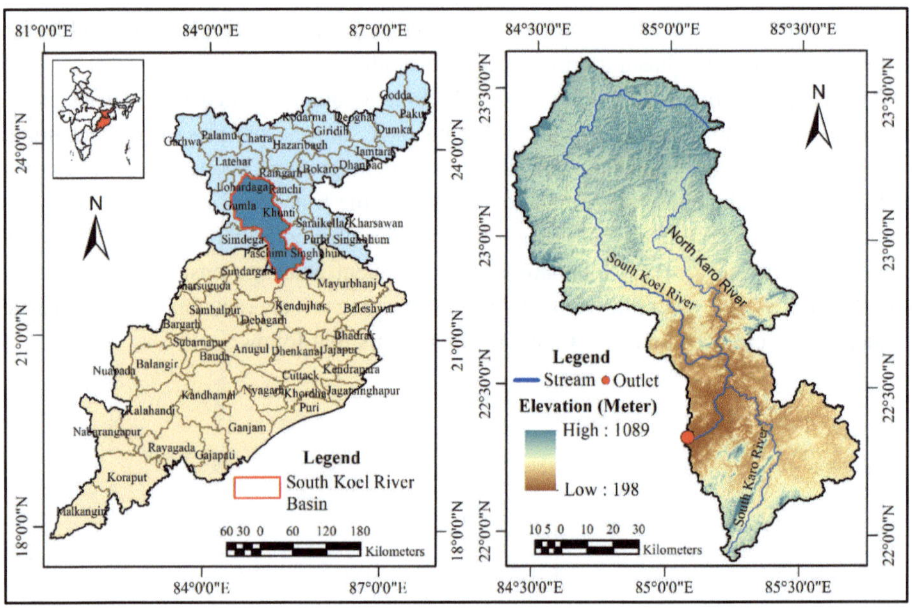

Fig. 3.1 Location of the study area. *Source* Census of India (2011) and SRTM DEM (2022)

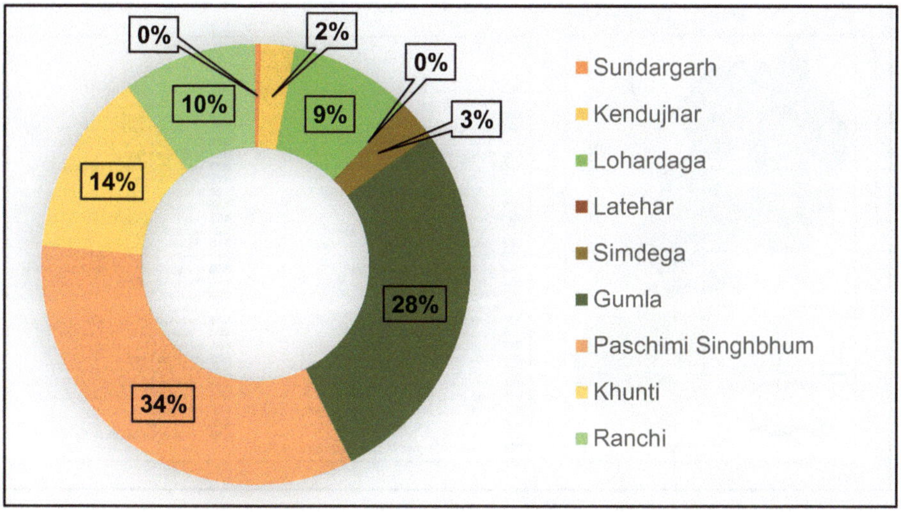

Fig. 3.2 District-wise areal coverage of the basin (%). *Source* Census of India (2011)

3.2.2 Geo-lithological Characteristics

The study area bears the imprints of various geological ages, with Proterozoic materials being the most dominant at 48.10% of the basin area, followed by Paleoproterozoic (29.85%), Archaean-Paleoproterozoic (12.76%), and Holocene (7.58%) materials. The Chhotanagpur gneissic complex is the most prominent geological formation, accounting for 49.51% of the area, whereas the Upper Bonai group, Lower Bonai group, and sediments make up the 14.08, 9.24, and 7.83% area of the basin respectively (GSI 2022). Granite and gneiss cover about 46.79% of the basin, while slate, phyllite, and mica schist together occupy 20.80%, and the remaining part is characterized by other rock types (GSI 2022) (Fig. 3.3).

3.2.3 Topography and Landforms

The basin's elevation varies between 198 and 1089 m (WGS84 datum), with an average elevation of 643.5 m. Situated on the fringes of the Chhotanagpur plateau, ~90% of the basin's area is located above the 300 m contour line. The northern section of the basin has higher elevations ranging from 569 to 1089 m, characterized by undulating terrain and heavily dissected regions. Similar elevation and topographic characteristics are also found in some parts of the southern portion of the basin. In contrast, the middle and most of the

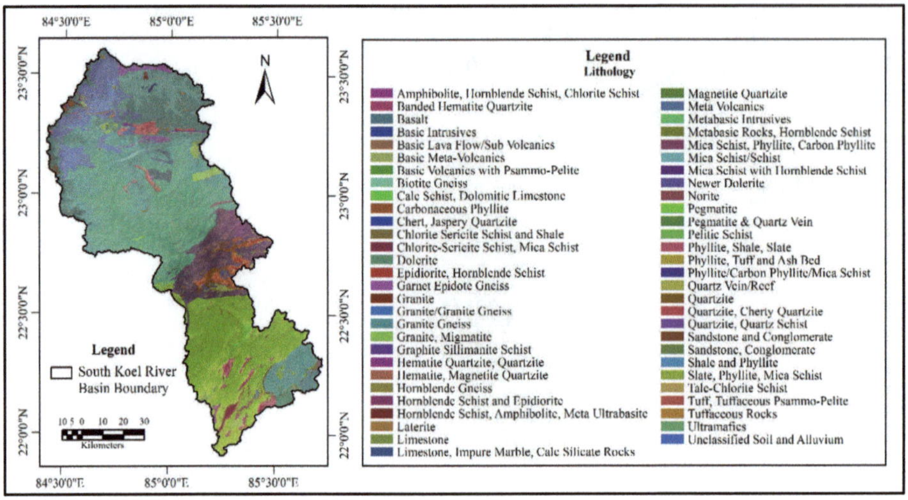

Fig. 3.3 Lithology of the study area. *Source* Geological Survey of India (2022)

southern portions have lower elevations ranging from 198 to 568 m. More than 64% of the basin's area consists of nearly level and gentle slopes (Fig. 3.4 and Table 3.1).

A pediment-peneplain complex, resulting from denudational activities, is the predominant landform type in the basin, covering about 62% of its area. In addition, the area has moderately dissected hills and valleys formed by both denudational and fluvial processes, each sharing about 15% of the basin. The denudational process also moderately dissected the upper plateau, which occupies 2.19% of the basin (GSI 2022). The distribution of geomorphic features is shown in Fig. 3.5.

3.2.4 The Hydrological Characteristics

South Koel is a perennial river having the highest stream flow in August (14,095 cumecs), followed by September (11,212 cumecs) and July (10,324 cumecs), while the lowest flow is observed in April (149 cumecs) (Fig. 3.6). Over the past 20 years, no floods have been recorded in this river basin. However, there have been some instances of water scarcity in certain years, namely 2001, 2004, 2005, 2010, 2012, 2015, and 2017 (Chaudhary and Pandey 2022).

In addition to that, since 2008–2009, a substantial portion of stream flow (1281 MCM) is getting diverted every year to the Subarnarekha River through a canal which accelerates the shortage of water resources (Sinha 2017).

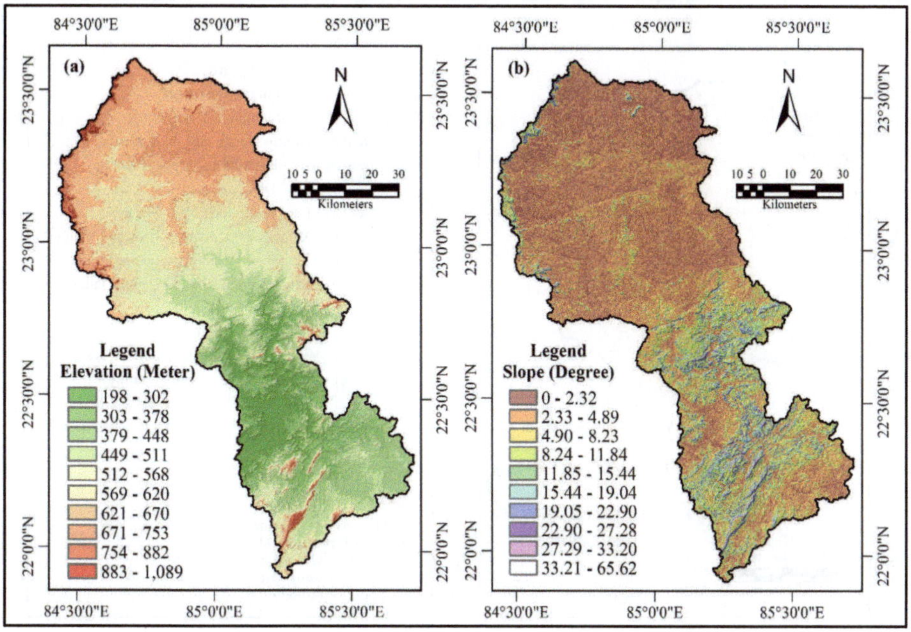

Fig. 3.4 **a** Elevation and **b** slope of the study area. *Source* SRTM DEM (2022)

Table 3.1 Elevation and slope zone of South Koel River Basin

Sl. No.	Elevation (m)	% of the basin area	Slope (degree)	% of the basin area
1.	198–302	6.54	0–2.32	32.81
2.	303–378	8.3	2.33–4.89	32.06
3.	379–448	11.07	4.90–8.23	14.5
4.	449–511	11.09	8.24–11.84	7.52
5.	512–568	15.36	11.85–15.44	5.07
6.	569–620	15.3	15.45–19.04	3.45
7.	621–670	16.45	19.05–22.90	2.37
8.	671–753	13.9	22.91–27.28	1.41
9.	754–882	1.34	27.29–33.20	0.65
10.	883–1089	0.65	33.21–65.62	0.16

Source Computed by the Authors from SRTM DEM (2022)

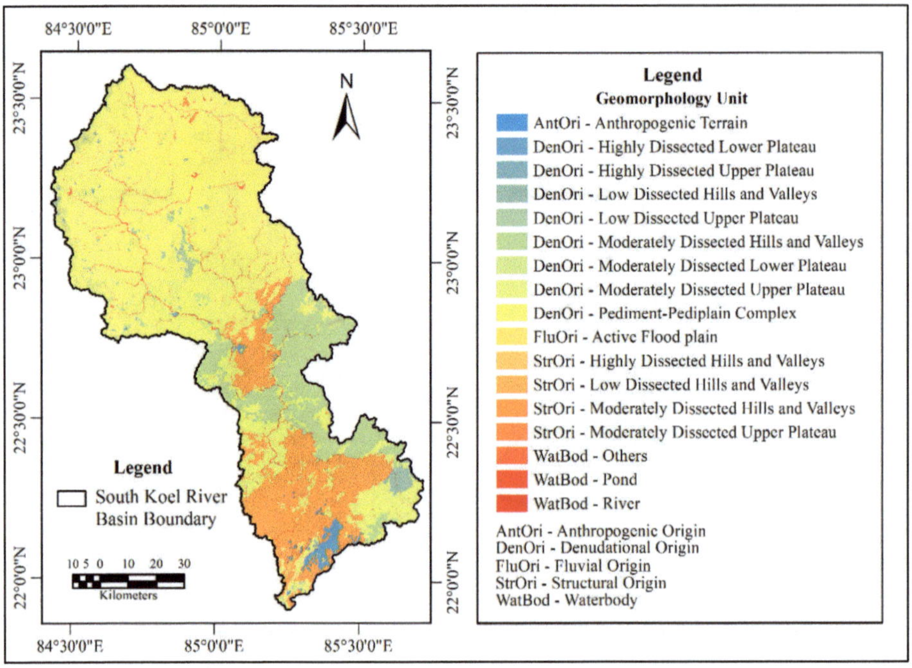

Fig. 3.5 Geomorphic features of the South Koel River basin. *Source* Geological Survey of India (2022)

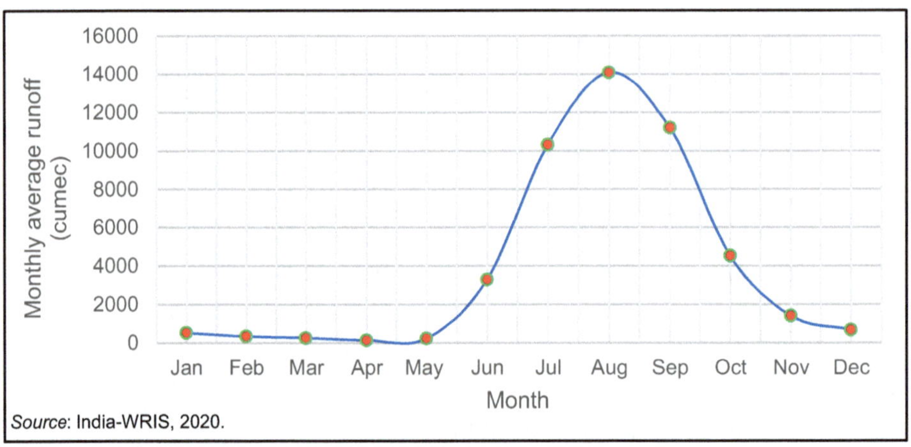

Fig. 3.6 Monthly average discharge of South Koel River at Jarikela (1981–2018)

3.2.5 Climate and Climate Variability

The basin experiences a tropical monsoon climate, characterized by distinct seasonal variations in temperature and rainfall. According to the IMD (cited in CWC 2014), the coldest months in the basin are December and January, with a minimum temperature of 12 °C. On the other hand, the hottest months are April and May, with maximum temperatures ranging from 35 to 38 °C. The mean annual maximum temperature in the basin is 37.67 °C, while the mean annual minimum temperature is 20.32 °C. The basin receives a mean annual rainfall of 1442.53 mm, with ~82% of it occurring during the rainy season (June–September). Around 90% of the total basin area experiences a mean annual rainfall between 1400 and 1600 mm (CWC 2014) (Fig. 3.7).

Tirkey et al. (2018) have reported that the basin is gradually experiencing by warming effect, and rainfall is likely to be decreasing, which is thought to have an adverse impact on stream flow and water resources. Guhathakurta and Rajeevan (2008) observed a decline in annual rainfall in Chhattisgarh, Jharkhand, and Kerala. Sharma and Singh (2017) also noted a significant decrease in annual, monsoon, and winter rainfall from 1901 to 2002, which could affect agricultural activities in Jharkhand. However, Praveen et al. (2020) found that while pre-monsoon and post-monsoon rainfall decreased, there was an increase in monsoon, winter, and annual rainfall during the period of 1901–2015, specifically in the state of Jharkhand.

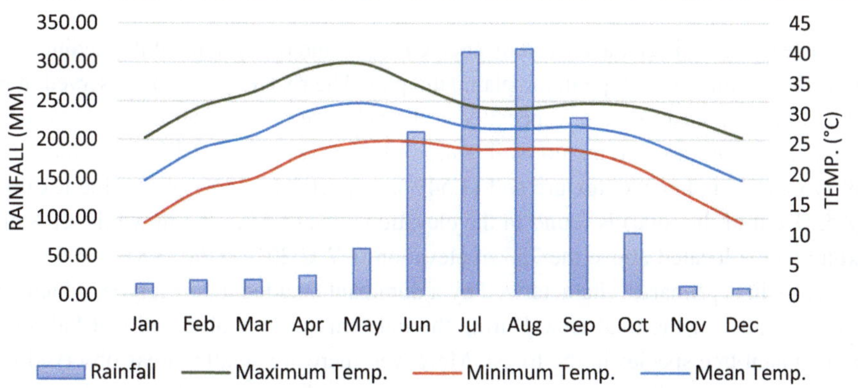

Source: Computed by the researcher from IMD gridded data.

Fig. 3.7 Mean monthly temperature and rainfall of the study area (1973–2018)

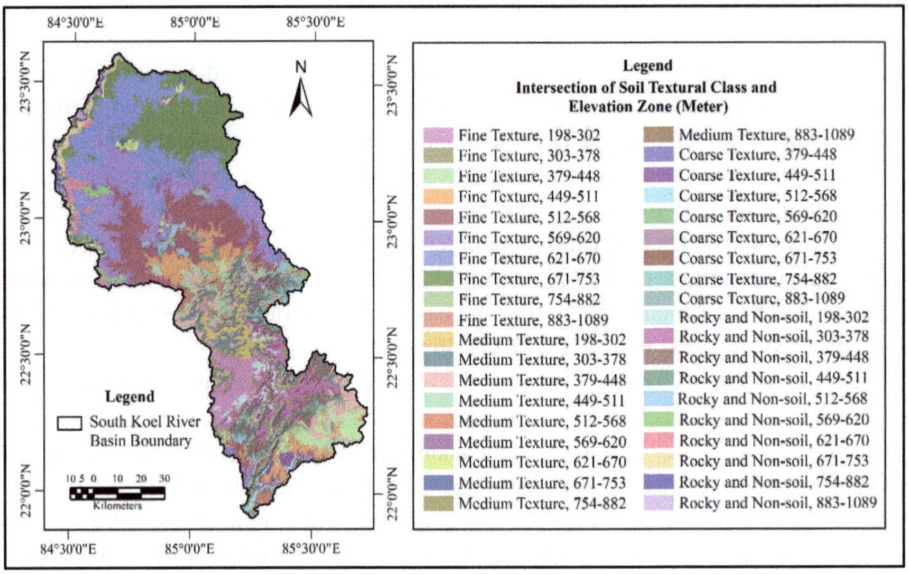

Fig. 3.8 Intersection of soil textural classes and elevation zones. *Source* Central Water Commission, Government of India 2014 and SRTM DEM (2022)

3.2.6 Soil and Vegetation

The Plateau region is characterized by coarse texture and rocky soil, while medium texture soil is the dominant soil type in the plain land part. The rock exposure is observed in many patches in the plateau region, especially in the northern part of the basin. About 63.45% area of the basin has fine texture soil, followed by rocky soil (19.57%), medium texture soil (15.44%), and coarse texture soil (1.54%), respectively. Most of the fine texture soil (78.55% out of the total) is found in the elevation zone of 512–753 m, while <1% of fine texture soil is located above the 753 m elevation (CWC 2014) (Fig. 3.8).

The basin is primarily characterized by a dominant deciduous forest type, which covers more than 30% of the total area. During the field survey, it is observed that Sal trees are the dominant tree species in the forest. Moreover, there are smaller areas observed within the basin during field visits that are covered by scrub forests and plantations.

3.2.7 LULC and Human Activities

As per the analysis of Landsat OLI satellite image, Nov 2016, and the subsequent ground truth verification, it is observed that agriculture constitutes the predominant land use in the basin, covering ~57.95% of its total area. The main crop cultivated in the area is paddy, which is grown once a year during the monsoon season from July to September. Forested areas encompass around 37.23% of the basin, while built-up areas account for 4.12%. Water bodies and bare ground cover 0.51 and 0.19% of the basin, respectively.

The highest population density (person/km^2) is found in Ranchi district (572), followed by Lohardaga (307), Kendujhar (217), Sundargarh (216), Khunti (210), Paschimi Singhbhum (208), Gumla (191), Latehar (169), and Simdega (159) districts respectively. Gumla and Lohardaga are the two notable urban centers in the basin, while there are many small towns like Sisai, Palkot, Kamdara, Ghagra, Basia (Gumla district), Kisko, Senha, Kuru (Lohardaga district), Mandar, Bero (Ranchi district), Khunti, Karra, Torpa (Khunti district), and Goilkera, Monaharpur, Noamundi, Jagannathpur (Paschimi Singhbhum) (Census of India 2011).

The majority of rural inhabitants in the basin are primarily engaged in agricultural activities, which are carried out exclusively only during the monsoon season. In the areas, on the other hand, wherever there are irrigation facilities, agriculture goes around the year. There are no significant large-scale industrial establishments in the basin. However, in the southern part of the basin, there are some mining areas.

Tirkey et al. (2018) have reported that the destruction of the forest to create new agricultural fields is the old tradition of this basin, while on the other hand, lack of water resources, some agricultural fields are turning into fallow and barren lands. So, an in-depth study is the need of the hour to assess the runoff impact of all these major changes in the basin so that accordingly remedial measures can be taken before the situation gets at its worst, and the present study is likely to address the issue.

3.2.8 Hydrologic Implications and Downstream Linkage

The interplay of hard-rock geology, steep-moderate slopes, monsoon-dominated rainfall and economic activities creates a runoff-dominated hydrologic regime in the South Koel Basin, characterized by limited groundwater storage and high flow variability. The basin's geology and topography promote rapid surface runoff with negligible infiltration (e.g., lack of deep groundwater retention), leading to flashy discharge dynamics. This combination controls both basin-scale discharge patterns and directly influences the Brahmani River's downstream contributions to the Brahmani-Baitarani delta.

The Brahmani–Baitarani delta forms a broad fan-shaped deltaic plain on Odisha's Utkal coastal plain, where the River Brahmani (formed by the combined flow of the South Koel and Sankh Rivers) and the River Baitarani deposit sediment before their outfall into the Bay of Bengal (Fig. 3.9). As a wave-influenced estuarine system, it develops distributary channels and mangrove-lined tidal creeks typical of a mixed fluvial-marine delta. The delta's hydrology is critically dependent on freshwater discharge and suspended sediment supply from upstream basins, including the South Koel watershed: reductions in

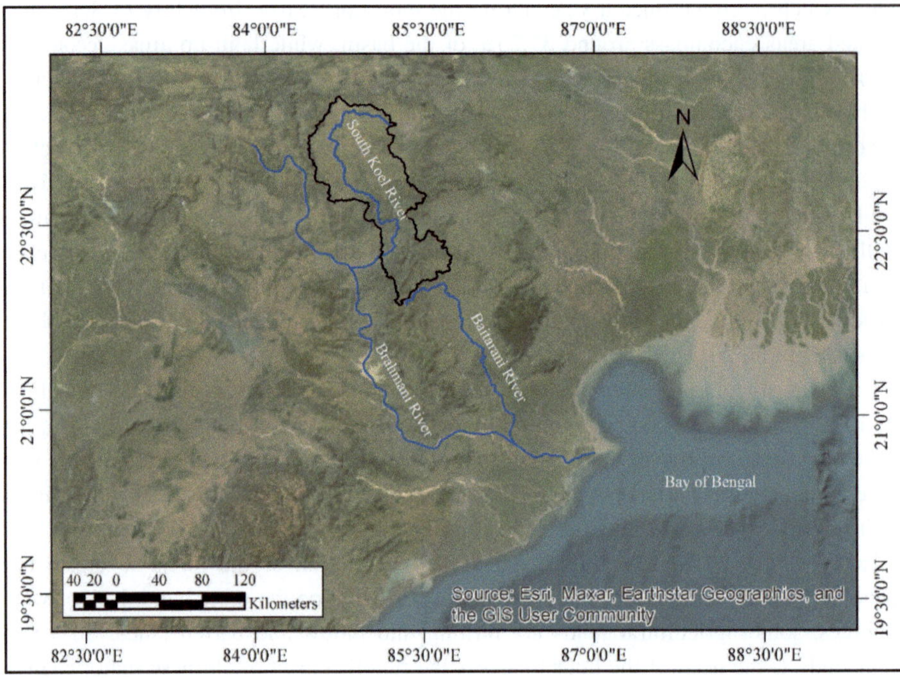

Fig. 3.9 Basin-to-delta connectivity

flow or sediment delivery upstream lead to diminished aggradation and deltaic land loss over time (Visakh and Raju 2018; Darby et al. 2020).

Maintaining regulated runoff and sediment load from the South Koel is essential for the delta's resilience. These inputs replenish tidal plains and support the Bhitarkanika, a mangrove ecosystem, which buffers coastal erosion, traps sediment, and interacts dynamically with saline tidal waters. Bhitarkanika is the second-largest mangrove forest ($\sim$650 km^2) in India, and a designated Ramsar wetland noted for its rich biodiversity, including estuarine crocodiles and sea turtles.

If upstream abstraction or land-use change reduces flow or sediment, the delta becomes more vulnerable to sea-level rise, saline intrusion, and ecological degradation of wetlands and fisheries (Darby et al. 2020). Thus, analyzing hydrologic transformations in the South Koel basin is fundamental for informing delta-scale management and maintaining the eco-geomorphic integrity of the Brahmani–Baitarani system.

3.3 Research Methodology

The present study employs a comprehensive methodological framework integrating multi-source datasets, geospatial analysis, statistical diagnostics, and hydrological modeling to investigate runoff alterations in the South Koel River Basin. The approach involves systematic data acquisition, pre-processing, analytical modeling, and validation

3.3.1 The Parameters and the Databases

The observations were made for the period between 1973 and 2018, and it was not possible to trace the phenomena prior to 1973 due to the absence of a reliable database for that period.

As per the *first objective* adopted for the present study, *rainfall* and *runoff* are the response variables which were analysed to detect step change and trend in the series against time in view to trace their temporal variability. So, it is a kind of univariate analysis.

As per the *second objective,* the only *response variable* is the *runoff,* whereas *climate* and *human activity* are the explanatory variables. The *rainfall, mean maximum* and *mean minimum temperature, humidity, solar radiation, and wind speed* were used as representative variables for *climate* and land use and land cover (LULC), on the other hand, was used as the only representative parameters for human activities.

As per the *third objective,* again, runoff is the only *response variable,* and three specific LULC classes, namely forest *cover*, *agricultural land use,* and *built-up area,* are the *explanatory variables.* The effect of rainfall, in this context, was controlled during the assessment of the functional relationship between response and explanatory variables. Here, two specific aspects of agricultural land use were taken into consideration: one was the *net shown area,* and the other was the *area-specific crop type* since this crop type has implications in the runoff generation mechanism. The database on crop type was generated by *primary survey* using the *questionnaire and schedule* in which stratified random sampling was used as the sampling technique, and the sample strata were generated based on elevation differences and soil types.

The Soil and Water Assessment Tool (SWAT), which is explained in detail in the fifth chapter, was used as a part of the whole data analysis method to achieve the second and third objectives. SWAT model requires four types of input data which include soil data, elevation data, LULC, and a weather database that include all the meteorological parameters as mentioned above, i.e., rainfall, maximum and minimum temperature, solar radiation, humidity, and wind speed. The discharge data was required for calibration and validation of the SWAT model.

For fulfilling the fourth objective i.e., to assess the influence of upstream hydrologic transformations in the South Koel Basin on downstream deltaic systems, the database on runoff trends, generated through time-series analysis and SWAT model outputs (1981–2018), served as the foundational input for analyzing upstream–downstream hydrologic connectivity. In addition to runoff, SWAT-simulated outputs on sediment yield and nitrogen transport, were critical for evaluating the fluvial contribution to delta morphology and ecological health. These outputs were derived from model simulations that had been calibrated and validated for the study period using observed hydrological data. The sediment and nutrient-related parameters were spatially and temporally analyzed to trace their delivery trajectories from the basin outlet toward the Brahmani–Baitarani delta. Thus, the combined use of runoff, sediment yield, and nitrogen transport datasets formed the core basis for quantifying the cascading impacts of basin-scale alterations on downstream deltaic resilience.

The sources and applications of all the databases used in the present study have been categorically mentioned in Table 3.2.

3.3.2 Data Pre-processing

The primary data on crop type were tabulated and cleaned properly in MS Excel. Similarly, the discharge and weather databases that were available on daily time steps from respective secondary sources were also cleaned properly and transformed into the desired format, e.g., daily to monthly or annual series, which were used as input at subsequent steps for model formulation and statistical analysis. The monthly time series of discharge data was used for validating the SWAT model, while the annual time series on the same parameter was used for statistical analysis. For SWAT run weather database was kept as a daily time series.

The missing value imputation was done by replacing the blank cell with an average value conditioned to corresponding points of time and places at which the observations were made. The data were checked for normality and autocorrelation in order to choose the required statistical tools for data analysis. Lag 1 autocorrelation was assessed by using Pearson's correlation coefficient, and normality was checked by using the Kolmogorov–Smirnov test and Shapiro–Wilk test. GIS data like DEM, Landsat TM/OLI, soil raster, etc., were processed accordingly in the GIS environment to make the data ready for use in the SWAT model.

3.3.3 Method of Data Analysis

First Objective: For step change analysis of rainfall and discharge series, the Buishand U test, the Buishand R test, and the Standard Normal Homogeneity test (SNHT) were applied if data were found to be normally distributed, otherwise, Pettitt's test and the Sequential Mann Kendall test were used. The last one can detect multiple change points,

Table 3.2 Sources and applications of all the databases used in the study area

Sl. No.	Databases	Sources	Applications	Software used
1.	Administrative, geological, lithological, and geomorphological maps	The administrative map was collected from the 2011 Census of India, and the remaining three maps were obtained from the Geological Survey of India 2022	To understand the basin's geophysical conditions	ArcMap 10.5
2.	Daily river discharge and reservoir discharge data	CWC, Department of water resources, River Development and Ganga Rejuvenation, Ministry of Jal Shakti, Govt. of India (India-WRIS 2020) (https://indiawris.gov.in/wris/#/)	The response variable in achieving all three objectives and also as input data for calibration and validation of the SWAT model	MS Excel 2021 RStudio v 1.2.5042 SPSS v.17.0 ArcSWAT 2012.10.5.24
3.	Daily weather data on maximum and minimum temperature, rainfall, solar radiation, humidity, and wind speed	NASA Power Data Access, Langley Research Center (NASA 2020) (https://power.larc.nasa.gov/data-access-viewer/)	Rainfall was used as the response variable in the first objective, while for the other two objectives, it was used as input data for calibration and validation of the SWAT model All other weather data was used as SWAT input data	MS Excel 2021 RStudio v 1.2.5042 SPSS v.17.0 ArcSWAT 2012.10.5.24
4.	Landsat TM and OLI image with 30 m resolution	USGS Earth Explorer, https://earthexplorer.usgs.gov (USGS Earth Explorer 2022)	To generate LULC, which was used as SWAT input data. The LULC map was available from NRSC and CWC, Govt. of India, and was also used as reference data for producing LULC maps from satellite images	QGIS 3.28.4 Erdas Imagine 2015 ArcMap 10.5 ArcSWAT 2012.10.5.24

(continued)

Table 3.2 (continued)

Sl. No.	Databases	Sources	Applications	Software used
5.	Soil grid data with 250 m spatial resolution and also soil map	Soil grid data from https://soilgrids.org (ISRIC-WDC 2020) and soil map from Brahmani Baitarani Basin Report (CWC 2014) Govt. of India	It was used as SWAT input data	ArcMap 10.5 ArcSWAT 2012.10.5.24
6.	SRTM DEM with 30 m resolution	https://earthexplorer.usgs.gov (USGS Earth Explorer 2022)	It was used as input data for the SWAT model	ArcMap 10.5 ArcSWAT 2012.10.5.24
7.	The area-specific crop type cultivated during monsoon seasons	Primary Survey through questionnaire and schedule using stratified random sampling	Explanatory variable to achieve the third objective	MS Excel 2021 ArcMap 10.5 ArcSWAT 2012.10.5.24
8.	Sediment and nutrient flux data	Output of SWAT simulation model	To understand impacts of basin-scale alterations on downstream deltaic system	ArcMap 10.5 ArcSWAT 2012.10.5.24

Source Compiled by the authors

whereas all others can detect only a single change point in the time series. It is usually recommended to apply more than one test so that one test result can be used to substantiate the other, and it is done in order to avoid the possibility of detecting any disguised break point in the series (Kundzewicz and Robson 2004). For trend analysis, the Mann–Kendall test and modified Mann–Kendall test were used for independent series and autocorrelated series, respectively. The magnitude of the trend was assessed by Sen's slope estimator (Fig. 3.10).

Second Objective: To assess the impact of climate variability and change in LULC on runoff alteration, the whole study period (1981–2018) was divided into two parts based on the result of step change analysis for runoff, and the earlier part was used as reference period while latter part as interference period.

The SWAT model was calibrated to LULC of both periods separately, and thereafter runoff was simulated separately using separate weather data sets of respective periods. The simulation of the reference period, here, is noted as 'Sim1' while the notation 'Sim2' is used to denote the simulation of the interference period. The mean difference between these two simulated runoff series, i.e., Sim2 and Sim1, is the total runoff alteration caused by both climate variability and human activity. Then, the model calibrated to reference

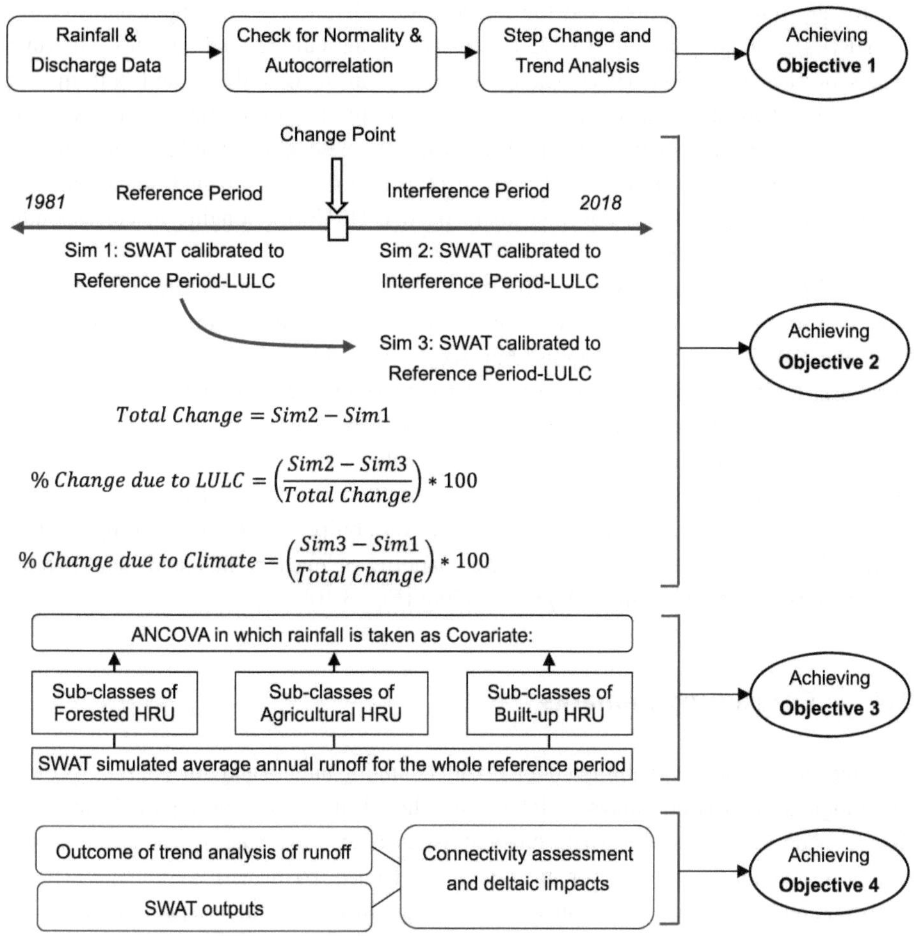

Fig. 3.10 Methodology of the study (Term 'Sim' stands for simulation)

period LULC was run using weather data of interference period, and this simulation is denoted by 'Sim3'. Now, the mean difference between Sim3 and Sim1 is the portion of runoff alteration caused by climate variation since LULC was the same for both simulations, but the weather dataset was different. On the other hand, the mean difference between Sim2 and Sim3 is the runoff alteration caused by human activity since, here, LULC is different, but the weather database is kept the same (Fig. 3.10).

Third Objective: To assess the runoff sensitivity to changing areal coverage of three specific individual LULC classes: forest cover, agriculture, and built-up area, the whole basin area was divided into a manageable number of homogeneous units technically called *Hydrologic Response Unit* (HRU) in such a way that each HRU contains only a single LULC class, i.e., either forest or agriculture or built-up, etc. Then the SWAT model was run for the reference period, and the average annual runoff was computed for each of the HRU(s) belonging to only LULC classes forest, agriculture, and built-up. The runoff values of each individual HRU belonging to the forest, agriculture, and built-up were further classified into four sub-categories based on the areal coverage of corresponding HRU(s) by using the quantile method and then three separate ANCOVA models with rainfall as a covariate were run for three respective LULC classes to analyse the sensitivity of runoff to changing areal coverage of respective HRU(s) belonging to corresponding LULC class (Fig. 3.10).

Fourth Objective: To analyze the upstream–downstream hydrologic linkages, the runoff reduction results were integrated with sediment yield and nitrogen transport data derived from the SWAT model simulations. For sediment yield analysis, the time-series data were first tested for autocorrelation to determine statistical dependencies across observations. Given the presence of autocorrelation, the Modified Mann–Kendall (MMK) Test was applied. This enabled the identification of statistically significant trends in sediment delivery over the 1981–2018 period. For nutrient transport assessment, SWAT-generated nitrogen data were used to examine temporal variability in nitrogen loading from the basin. The outputs were analyzed to infer patterns of nutrient flux and their potential implications for downstream ecological loading (Fig. 3.10).

3.4 Concluding Remarks

The impact of climate variability coupled with changing human activities in the river basin hydrologic system is nowadays a global issue though it has its own regional implication due to wide variations in surface characteristics across the basins. So, observation in one basin cannot be inferred for another basin. Hence, basin-wise systematic scientific investigation is of utmost importance in view to get insight into this aspect in the regional context.

The present study in South Koel River Basin has been formulated in view to address this global issue but in the regional context. The methodology adopted here, including trend analysis of rainfall-runoff, application of the SWAT model, integration of LULC and climate datasets, and analysis of sediment and nutrient fluxes, has already been validated in many previous works, and in the present study, it has been designed carefully to make it more relevant to the existing geophysical characteristics of the selected basin. So, the findings of current research are expected to be reliable outcomes that can be an effective contribution to the discipline.

4 Temporal Variability of Rainfall and Runoff

4.1 Prelude

The first objective of the present study is to assess the temporal variability of rainfall and runoff since 1973 in terms of trend and step change in the South Koel River Basin. A trend represents a gradual, systematic change in a variable over time, typically observed as a consistent increase or decrease in the series. It is one of the four structural components of a climate variable, alongside seasonality, cyclicity, and irregularity. A step change, by contrast, signifies an abrupt shift in the mean level of a time series, creating a structural break that divides the record into distinct sub-periods. It is often linked to climatic regime shifts (sudden reorganization of the climate system), or caused by changes in measurement practices, station relocation, or human interventions.

Both of these features are widely regarded as key indicators of climate change because they directly capture long-term or regime-level transformations in climate conditions. A statistically significant upward or downward trend, for example, provides clear evidence of gradual and sustained changes such as global warming or cooling, which are the central signatures of climate change. Similarly, the detection of a step change indicates an abrupt shift in the baseline state of the system, often linked to external forcing (e.g., greenhouse gas emissions, volcanic eruptions) or internal reorganizations of large-scale circulation patterns, and thus provides another critical signal of changing climate.

In contrast, other components of a time series, such as seasonality, cyclicity, and random variations, primarily represent natural variability inherent in the climate system. Seasonal cycles are regular and expected annual fluctuations, while cyclicity reflects oscillations such as ENSO or Pacific Decadal Oscillation (PDO) that occur over interannual to decadal scales, and random components capture unpredictable short-term noise. While these features are important for understanding climate variability, they do not serve as robust markers of long-term climate change.

S. Mondal and R. K. Paul, *Climate Variability and Human Interference*, Deltas of the World, https://doi.org/10.1007/978-3-032-13257-4_4

For this reason, trend and step-change analyses play a central role in climate change detection, as they reveal enduring or abrupt departures from the historical baseline of the climate system. In this chapter, these analyses were applied to long-term rainfall and runoff series in the South Koel River Basin to examine the temporal variability of the series and fulfill the first objective of the study.

4.2 Database and Methods

4.2.1 Data Sources

The daily discharge data (m^3/s) were collected from the Department of water resources, River Development and Ganga Rejuvenation, Ministry of Jal Shakti, Govt. of India (https://indiawris.gov.in/wris/#/). This data is available for the period between 1973 and 2018 for the gauging station at Jarikela (Fig. 4.1), which is the most downstream station of the South Koel River basin, having an availability of discharge data.

On the other hand, the daily rainfall gridded data available from the NASA power data access, Langley Research Centre (https://power.larc.nasa.gov/data-access-viewer/) were

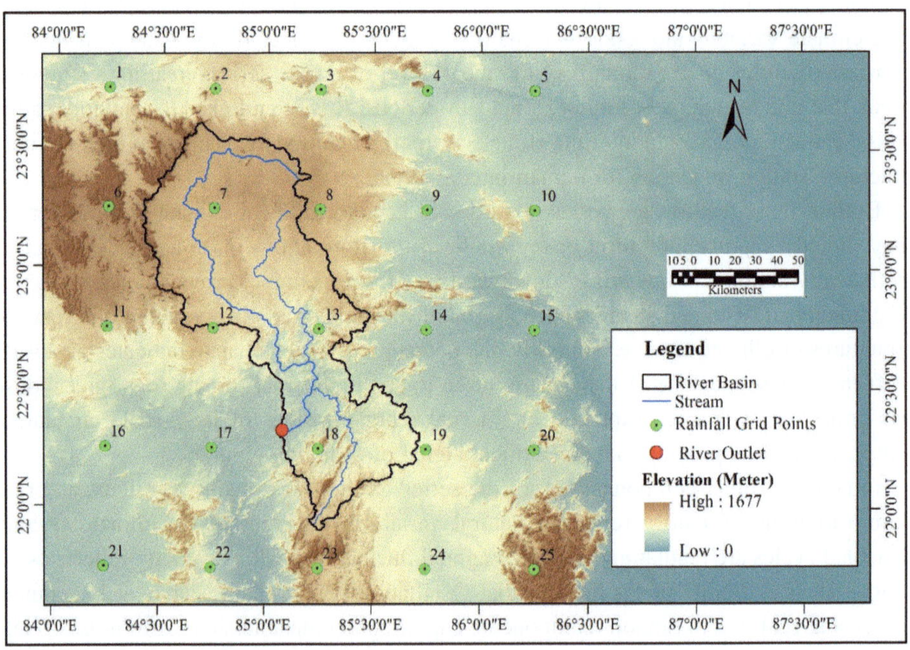

Fig. 4.1 Distribution of rainfall grid points and the outlet of the basin. *Source* SRTM DEM (2022) and NASA Power data access (2020)

used in the present study. The data are available since 1981 at a grid resolution of 0.625° × 0.50°.

The present analysis could have been done by using IMD gridded data which are available at a finer resolution (0.25° × 0.25°) from India Meteorological Department (IMD), but it was not done so. It is because IMD gridded data provides only daily maximum temperature and daily minimum temperature along with daily rainfall. On the other hand, apart from these three, a set of another three weather parameters, namely humidity, solar radiation, and wind speed, are also required to run the SWAT simulation model, which was applied to achieve the second objective as discussed in next chapter, but these three datasets are not available from IMD, whereas NASA power access is the single source which provides all six weather parameters required in the present study and it is highly recommended to use the weather datasets from a single source instead of multiple sources to simulate the runoff in order to reduce the uncertainty in hydrological modelling and improve the accuracy of runoff simulations (Guo et al. 2004).

The SWAT website, on the other hand, also provides all the required six input weather datasets at a grid resolution of 0.25° × 0.20°, but they are available up to 31/07/2014, whereas this study needs the data as updated as possible. For these reasons, the NASA Power dataset was used in the present study.

4.2.2 Data Pre-processing

(a) *Data Tidying*

It was observed that all the datasets were already cleaned properly except for the presence of two missing frequencies in the discharge series for the date of 30/09/2017 and 22/12/2017. The missing value imputation was done by replacing the blank cells with the arithmetic average value conditioned to specific points of time and places at which the observations were made (see Sect. 3.2).

(b) *Parameter Computation*

After the necessary corrections, the daily discharge data (m^3/s) were transformed into daily *runoff depth* value (mm), and then based on runoff depth, the *year-wise seasonal and annual runoff total* (mm) were computed for the whole basin for the duration between 1973 and 2018. The daily rainfall data (mm), on the other hand, were directly transformed into *year-wise total seasonal and annual rainfall series* for the period between 1981 and 2018. These parameters were computed as the spatial average value for the whole basin as well as the series for respective grid points (Table 4.1).

It is important to note that discharge data have been available since 1973; accordingly, the runoff series was constructed for the period 1973–2018 to address the first research

Table 4.1 The variables generated based on available discharge and rainfall data for further analysis in view to achieve the first research objective

Parameters	Specific variables generated	Series length	Remarks
Runoff (mm)*	Year-wise pre-monsoon runoff total	The series was generated for the following two periods (i) 1973–2018** (ii) 1981–2018	(i) The runoff (mm) was generated as the spatial average of the whole basin (ii) The rainfall (mm) series was generated for the individual grid point and as the spatial average for the entire basin. The spatial average rainfall for the whole basin was computed by using the Thiessen Polygon method
	Year-wise monsoon runoff total		
	Year-wise post-monsoon runoff total		
	Year-wise winter runoff total		
	Annual runoff total		
Rainfall (mm)	Year-wise pre-monsoon rainfall total	The series were generated for the period 1981–2018	
	Year-wise monsoon rainfall total		
	Year-wise post-monsoon rainfall total		
	Year-wise winter rainfall total		
	Annual rainfall total		

*The daily discharge values (m^3/s) were transformed into runoff depth (mm) for the whole basin, and then based on this value, the concerned parameters were generated
**The discharge data were available since 1973, and hence the runoff series was generated for the period 1973–2018 as per the first research objective. Again, the runoff data was also generated for the period 1981–2018 to make it comparable with the rainfall series which is available since 1981. Both the variables need to be pertaining to the same period for impact analysis which was done in order to achieve the second research objective, which is discussed in Chap. 3
Source Computed by the authors

objective. For comparability with rainfall data, however-available only from 1981-the runoff series was also generated for 1981–2018. Aligning both variables over a common time frame is essential for the impact analysis required to fulfill the second research objective.

(c) *Exploratory Data Analysis*

All the generated variables were checked for normality and autocorrelation so that suitable analysis tools could be chosen accordingly.

(i) *Check for Normality*

The normality was tested by using the Kolmogorov–Smirnov test and the Shapiro–Wilk test. Whenever they gave conflicting outcomes, the Kolmogorov–Smirnov test result was accepted since the concerned test is used for a series longer than 50 observations which happen to be true in the present case. Both the tools are described below.

(1) *Kolmogorov–Smirnov Test:*

The Kolmogorov–Smirnov test, developed by A. K. Kolmogorov and N. V. Smirnov in the 1930s as cited by D'Agostino and Stephens (1986), is a widely used non-parametric statistical method for assessing the similarity between two probability distributions or testing the fit of a sample to a specific distribution. It is generally used when the sample is more than 50. The test measures the maximum difference between the empirical distribution function $F_0(x)$ of the observed data and the cumulative distribution function $F_r(x)$ of the reference distribution which is referred to as the D-statistic. The formula for computing D-statistic is as follows:

$$D = \max|F_0(x) - F_r(x)| \tag{4.1}$$

The significance of the *D*-statistic is determined using critical values or algorithms, which can be obtained from tables or assessed through resampling techniques (Alice Robson et al. 2000). The assumption of normality was rejected when the *p*-value of the D-statistic was found to be ≤ 0.05.

(2) *Shapiro–Wilk Test:*

The Shapiro–Wilk test is a non-parametric test that was used as an alternative to the Kolmogorov–Smirnov test to check how close the sample data is fitted to a normal distribution. It is generally used when the sample size is <50, although it is also being used for large sample datasets with equal importance (Mishra et al. 2019).

The test statistics, known as W-statistics, represent the difference between the theoretical frequency distribution and observed frequency distribution. First, the data series are sorted in ascending order, i.e., $x_1 \leq \ldots \leq x_n$, and then W statistics are calculated using the following formula:

$$W = \frac{\left(\sum_{i=1}^{n} a_i x_{(i)}\right)^2}{\left(\sum_{i=1}^{n} (x_i - \overline{x})\right)^2} \tag{4.2}$$

Coefficients a_i has been taken from the Shapiro–Wilk table, and the p-value was calculated using a linear ratio (Shapiro and Wilk 1965). The null Hypothesis (H_0) states that data is normally distributed.

(ii) *Check for Autocorrelation*

Before performing trend analysis of rainfall and runoff data series, it is also compulsory to test the independence of the data series against autocorrelation since many trend analysis tools have an assumption of independence. In the present study, as suggested by Brockwell and Davis (2016), lag-1 autocorrelation for each series was calculated using Pearson's product-moment correlation, and its significance was assessed using a t-test. The null hypothesis was rejected at the 5% significance level.

4.2.3 Method of Data Analysis

As per the first objective of the study, two types of analysis were done: one is the analysis of *step change* (also called *sudden change*), and another is the *trend*. The statistical tools that were used for both analyses are described below separately. However, all the computations were done using RStudio v. 1.2.5042.

(a) *Methods of Step Change* Analysis

Step change analysis, also referred to as *change point analysis*, involves detecting and describing notable shifts or sudden changes in the behaviour or pattern of a time series data based on the changes in mean, variance, or other relevant characteristics (e.g., Median).

In the present study, *Pettitt's test* was applied for detecting change point for the series which does not follow the normal distribution, whereas for the normally distributed series *Buishand U test*, *Buishand Range test*, and *Standard Normal Homogeneity test* (SNHT) were applied. All these tools are used for detecting single change point in the time series, whereas an attempt was also made for detecting multiple change points, and for this purpose, the Sequential Mann-Kendall test was used. All these tools are described below.

(i) Pettitt's Test

Pettitt's test (or median change point test) is a non-parametric statistical test applied to identify a single change point in a time series dataset (Pettitt 1979; Alice Robson et al. 2000). This test was based on the Mann–Whitney test and did not rely on assumptions about the distribution of the data, making it suitable for use in situations where data distribution is uncertain (Pettitt 1979; Pérez et al. 2011). It is particularly effective for large sample sizes. The test statistic, denoted by U*, is defined for the data series. $x_i \ldots x_N$ as follows:

$$U_{t,N} = \sum_{i=1}^{t} \sum_{j=t+1}^{N} sign(x_i - x_j) \tag{4.3}$$

where the sign function donated by "sign, assigns the values 1 for "positive", 0 for "zero", and –1 for "negative" difference. The calculation of variable $U_{t,N}$ involves the cumulative sum in the summation (for $i = 1, 2, \ldots, t$). The statistic is comparable to the Mann–Whitney statistic, which is used to test whether two samples. $x_i, \ldots, x_t$ and $x_{t+1}, \ldots, x_N$ originate from the same population (Pettitt 1979). It was assessed for various values of t with $1 \leq t < N$ (Pettitt 1979). The null hypothesis (H_0) assumes differences in the median of the data series, while the alternative hypothesis (H_1) suggests the presence of a change. The following test statistic, as proposed by (Pettitt 1979), was employed to evaluate the null hypothesis.

$$K_N = \max_{1 \leq t < N} \left| U_{t,N} \right| \tag{4.4}$$

The series exhibits a single change point at time t, indicated by the observation K_N (Pérez et al. 2011). Pettitt (1979) proposed the following statistics to identify changes that occur in one direction and have the same distribution under the null hypothesis.

$$\text{For a positive change, } K_N^+ = \max_{1 \leq t < N} U_{t,N} \tag{4.5}$$

$$\text{For a negative change, } K_N^- = \min_{1 \leq t < N} U_{t,N} \tag{4.6}$$

Pettitt (1979) showed that the rank statistics of variable $U_{t,N}$ can be calculated based on the rank theory. Assuming the null hypothesis, with k representing the K_N value of the series, the variable is computed as follows:

$$U_{t,N} = 2\sum_{i=1}^{k} r_i - k(N+1) \tag{4.7}$$

where k takes values from 1 to N, and $\sum r_i$ represents the cumulative sum of the ranks assigned to the original values. The ranks were assigned in ascending order, where the smallest value in the series is assigned rank 1, and the Nth observation is assigned rank r_N. Ties were handled similarly to how they are addressed in the computation of the Mann–Whitney statistic or Spearman's rank correlation coefficient.

The exceeding probability of a given value of k was calculated in the following manner (Pettitt 1979; Pérez et al. 2011):

$$\text{Prob}(K_N > k) \approx 2\exp\left[\frac{-6k^2}{N^3 - N}\right] \tag{4.8}$$

Alice Robson et al. (2000) utilized Bernoulli experiments for probability calculation. The null hypothesis (H_0) is rejected when the estimated probability is below the significance level α. They proposed a modified version of the test using a different test statistic K_N^* as an alternative indicator for detecting change points. The formula is as follows:

$$K_N^* = \max\left|\frac{U_{t,N}}{\sqrt{Nt - t^2}}\right| \tag{4.9}$$

Resampling methods should be employed to estimate the significance level for the above test statistic (Alice Robson et al. 2000).

(ii) *Sequential Mann–Kendall Test*

The Sequential Mann–Kendall test or plot (also called Sequential Mann–Kendall–Seyer test) is a non-parametric method designed to detect potential turning points in a time series (Kendall and Gibbons 1990). It can identify multiple change points and assumes that the time series x is equidistant. The test calculates two sets of Kendall's tau values, one in a progressive order and the other in a retrograde order. Intersection points of these two lines are considered approximate potential turning points in the trend (Sneyers 1990). If the values in either the progressive or retrograde row exceed specific confidence limits both before and after the intersection points, the trend turning point is considered statistically significant at the corresponding confidence level of 1.96 for a 95% confidence level.

The test compares ranked values y_i from the original dataset $(x_1, x_2, x_3 \ldots x_n)$. For each comparison, instances where $y_i > y_j$ are counted and represented as n_i. A test statistic t_i is then defined based on these counts, as shown below.

$$t_i = \sum_{j=1}^{i} n_i \tag{4.10}$$

The mean of the distribution of t_i is represented as follows:

$$E(t_i) = \frac{i(i-1)}{4} \quad (4.11)$$

Similarly, the variance of t_i is expressed as follows:

$$Var(t_i) = \frac{i(i-1)(2i+5)}{72} \quad (4.12)$$

The reduced or standardized variable (t_i) was computed for each of the test statistic variables t_i is presented as follows:

$$u(t_i) = \frac{[t_i - E(t_i)}{\sqrt{\mathrm{var}(t_i)}} \quad (4.13)$$

The forward sequential statistic, denoted as $u(t_i)$, was computed using the original time series $(x_1, x_2, x_3 \ldots x_n)$. Conversely, backward sequential statistic, denoted as $u'^{(t_i)}$, was computed by sorting the time series in reverse order. By plotting the curves of $u(t_i)$ and $u'(t_i)$ and examining their intersections, potential trend turning points can be identified. If the intersection occurs beyond ± 1.96 (at the 5% level) of the standardized statistic, it indicates a detectable change in the data series at that point. Furthermore, if any value of the reduced variable exceeds a predetermined significance level determined by a Gaussian distribution, it becomes possible to reject the null hypothesis (H_0: the analyzed sample does not exhibit the initiation of a new trend). In this study, simultaneous change points of related components approach based on the Sequential Mann–Kendall test were used to determine the change points in the data series as proposed by Dey and Mishra (2017).

(iii) *Buishand U Test*

The Buishand U test, a parametric statistical method given by Buishand (1981, 1982, 1984), is used to identify the specific point of time in which the mean of an annual time series undergoes a shift. This test assumes the series follows a normal distribution and is independent (Wijngaard et al. 2003; Yozgatligil and Yazici 2016). It uses adjusted partial sums to detect breaks occurring within the time series, demonstrating higher sensitivity to such breaks (Hawkins 1977; Wijngaard et al. 2003). The adjusted partial sum is calculated based on the mean value of $\overline{Y}$ of the series $Y_1, \ldots, Y_n$ (Buishand 1982; Hänsel et al. 2016) by using the formula as given below:

$$S_0^* = 0, \text{ and } S_k^* = \sum_{i=1}^{k} (Y_i - \overline{Y}), \quad k = 1, \ldots, n \quad (4.14)$$

The term 'summation' (for $i = 1, 2 \ldots, k$) used in this context refers to the process of calculating the cumulative sum of a variable S_k^*. When $k = n$, and $S_n^* = 0$, the variable S_k^* is applied to a homogeneous series in which the components of the series deviate randomly from their mean without any consistent pattern, the variable will exhibit fluctuations around zero. However, if there is a break in the series at a particular year (K), the S_k^* variable will reach either a maximum (if there is a negative shift) or a minimum (if there is a positive shift) near that year $k = K$ (Yozgatligil and Yazici 2016). To test for statistical significance, Buishand (1982) developed three test statistics as given below:

$$Q = \max_{0 \le k \le n} \left| \frac{S_k^*}{D_Y} \right|; \tag{4.15}$$

$$R = \left[\max_{0 \le k \le n} S_k^* - \min_{0 \le k \le n} S_k^* \right] \Big/ D_Y \tag{4.16}$$

$$U = \frac{1}{n(n+1)} \sum_{k=1}^{n-1} \left(\frac{S_k^*}{D_Y} \right)^2 \tag{4.17}$$

The standardized variable S_k^* is obtained by dividing it by the standard deviation D_Y of the series. Three test statistics, namely the Q-statistic (maximum value of rescaled adjusted partial sum), R-statistic (rescaled adjusted range), and U-statistic (based on Gardner's original formulation assuming known variance) are calculated using this standardization (Pérez et al. 2011). In the case of unknown variance, the sample variance D_Y^2 is used to calculate the U-statistic (Buishand 1982). The Q-statistic indicates the presence of a break at time k in the series.

The null hypothesis (H_0) presupposes that the series is homogeneous in nature, while the alternative hypothesis (H_1) suggests a break in the series. Buishand (1982) stated that a large value of the test statistic indicates the presence of a break. The critical value for the test statistics, namely $\frac{Q}{\sqrt{n}}$, $\frac{R}{\sqrt{n}}$ and U, are provided by Buishand (1982). For $n \to \infty$, the critical value of Q can be determined using a table that is specifically designed for the Kolmogorov–Smirnov goodness-of-fit statistic (Buishand 1982).

(iv) *Buishand Range Test*

The Buishand Range test is another parametric statistical method given by Buishand (1981, 1982 , 1984) to detect a change point of a normal variate. In this test, the rescaled adjusted partial sums were calculated using Eq. 2.14, and the test statistic $\frac{R}{\sqrt{n}}$ which is defined by using Eq. 2.16 as given by Buishand (1982). The *p*-value of $\frac{R}{\sqrt{n}}$ is calculated using a Monte Carlo simulation method using m replicates. The critical values are determined based on m = 19,999 Monte Carlo simulations, as tabulated by Buishand (1982).

(v) *Standard Normal* Homogeneity Test *(SNHT)*

Alexandersson (1986) developed a parametric statistical test to identify a single change point in annual precipitation data which is based on two assumptions: the data follows a normal distribution, and there is only one change point resulting in a shift in the mean value. The null hypothesis (H_0) assumes a homogeneous distribution with a mean of zero and a unit standard deviation. Conversely, the alternative hypothesis (H_1) proposes a single break at some point between the first and last observation (*k* must be between 1 and *n*, where $1 \leq k \leq n$). According to Alexandersson (1986), if there is a single change point in the data series, the mean before the break (μ_1) is different from the mean after the break (μ_2), i.e., $\mu_1 \neq \mu_2$ of the data series.

Given a set of annual observations $x_i \ldots x_n$ (for $i = 1, 2, 3 \ldots n$)with a mean of $\overline{X}$ and a standard deviation of S_X, the test statistic for assessing the null hypothesis can be computed using the following formula.

$$T_0 = \max_{1 \leq k \leq n}\{T_k\} \tag{4.18}$$

$$T_k = k\overline{z}_1^2 + (n-k)\overline{z}_2^2, \quad \text{for } k = 1, 2 \ldots n \tag{4.19}$$

$$\overline{z}_1 = \frac{1}{k}\sum_{i=1}^{k}\left(\frac{X_i - \overline{X}}{s_X}\right) \text{ and } \overline{z}_2 = \frac{1}{n-k}\sum_{i=k+1}^{n}\left(\frac{X_i - \overline{X}}{s_X}\right) \tag{4.20}$$

The symbol Σ, used in the formula to create the variables $\overline{z}_1$ and $\overline{z}_2$, represents the sum of all values in a sequence, where *k* ranges from 1 to *n* (for $k = 1, 2 \ldots n$). In other words, Σ indicates a cumulative sum of values.

If the calculated value of the test statistic T_0 exceeds the critical value, the null hypothesis is rejected. The critical values for various sample sizes between 10 and 5000 can be obtained from the study conducted by Khaliq and Ouarda (2007) and Patakamuri et al. (2020). Alexandersson (1986) mentioned that the actual data series was substituted with a series of ratios $|q_i|_{i=1}^{n}$ obtained by comparing the observed values with those from a reference station. This substitution aims to reduce noise in the observed data series (Alexandersson 1986; Pérez et al. 2011).

(b) *Methods* of Trend Analysis

In simple terms, a *trend* as described earlier, refers to the process of gradual increase or decrease of data series against time. In this study, rainfall and runoff were the response variables which were analysed with respect to the time by applying the Mann–Kendall test for the series, which has no autocorrelation, and by applying the Modified Mann–Kendall test for the autocorrelated series. The computation in the present study was done by using RStudio v. 1.2.5042. However, both methods are described below.

(i) *Mann–Kendall Test*

The Mann–Kendall test (or Kendall *t*-test) is a non-parametric test that is widely used and recommended by the World Meteorological Organization (WMO) for examining trends in the meteorological time series data (Ali et al. 2019). It is a special case of the Kendall rank correlation test and was introduced by Hirsh et al. (1982) based on Mann's (1945) original work (Mann 1945; Kendall 1975; Hirsch et al. 1982; Hipel and McLeod 1994). The test is designed to assess the randomness over time using Kendall's rank correlation method (Kendall 1975), and it is comparable to the significance tests used for simple linear regression and Pearson's correlation coefficient (Helsel et al. 2020).

The null hypothesis (H_0) assumes that the data originated from a population where the random variables are independent and follow the same distribution. Conversely, the alternative hypothesis (H_1) proposes the presence of a monotonic trend in the data over time (Hipel and McLeod 1994). The test statistic which is employed to assess the H_0 hypothesis is computed using the following equation:

$$S = \sum_{k=1}^{n-1} \sum_{j=k+1}^{n} \operatorname{sgn}(x_j - x_k), \quad \text{Where,} \quad \operatorname{sgn}(\theta) = \begin{pmatrix} +1 \; if \; \theta > 0 \\ 0 \; if \; \theta = 0 \\ -1 \; if \; \theta < 0 \end{pmatrix} \tag{4.21}$$

Kendall (1975) demonstrated that when the sample size exceeds 10, the variable S follows a normal distribution, and its variance is computed by the following formula:

$$Var(S) = [n(n-1)(2n+5) - \sum_{j=1}^{P} t_j(t_j-1)(2t_j+5)]/18 \tag{4.22}$$

The formula considers the presence of tied groups in the data, represented by the number of tied groups (P) and the number of data points (t_j) within each group (jth). If there are no tied data points, the formula is reduced to $[n\ (n-1)\ (2n+5)]$. For small sample sizes, Mann (1945) and Kendall (1975) computed the distribution of S, while for larger sample sizes, the normal distribution approximation is used (Hipel and McLeod

1994), and in that case, the test statistic Z is computed using the following equation:

$$Z = \begin{pmatrix} \frac{S-1}{[Var(S)]^{1/2}} \; if \; S > 0 \\ 0 \qquad\quad if \; S = 0 \\ \frac{S+1}{[Var(S)]^{1/2}} \; if \; S > 0 \end{pmatrix} \tag{4.23}$$

A positive value of test statistic shows an increasing trend over time, while a negative value suggests a declining trend (Hirsch et al. 1982). The Sen's Slope, originally proposed by Sen (1968), was used here to estimate the magnitude of the trend. For all $k \leq j$, and $i = 1, 2, ..., N.$., Sen's slope (β_i) is defined as the median value of following estimate (Patakamuri et al. 2020):

$$\beta_i = \frac{x_j - x_k}{j - k} \tag{4.24}$$

(ii) *Modified Mann–Kendall Test*

In datasets where autocorrelation is present, the Modified Mann–Kendall test is employed. Autocorrelation can influence the detection of trends by increasing the likelihood of detecting trends even in the absence of actual trends in the data (positive autocorrelation) or by reducing the likelihood of identifying trends even when they actually exist (negative autocorrelation) (Hamed and Ramachandra Rao 1998).

To address a limitation in the original Mann–Kendall test regarding the variance of the test statistic, Hamed and Ramachandra Rao (1998) proposed a modification. This modification introduced a revised variance formula for computing the variance of test statistic S. The modified formula is as follows:

$$V^*(S) = \text{var}(S) * \frac{n}{n_s^*} \tag{4.25}$$

$$\frac{n}{n_s^*} = 1 + \frac{2}{n(n-1)(n-2)} * \sum_{i=1}^{n-1} (n-i)(n-i-1)(n-i-2)\rho_s(i) \tag{4.26}$$

In the above equations, a correction factor n/n_s^* is incorporated to address the impact of autocorrelation in the dataset. The correction factor considers the total number of observations (n) and the autocorrelation function of the ranks of the observations denoted as $\rho_s(i)$ (Hamed and Ramachandra Rao 1998; Patakamuri and O'Brien 2020). The Modified Mann–Kendall test is more suitable than the original version as it retains its effectiveness in detecting trends. Therefore, in this study, the data were examined for autocorrelation, and the statistical tool was chosen accordingly to capture the trends in the data series.

4.3 Results and Discussion

4.3.1 The Results of Exploratory Data Analysis (EDA)

(a) *Result of Normality Test*

The result of the Kolmogorov–Smirnov test and that of the Shapiro–Wilk test is found to be consistent with each other for year-wise seasonal and annual rainfall (mm) and runoff (mm) total for entire basin during 1981–2018 (Table 4.2). All the seasonal and annual rainfall series are found to be normally distributed except the post-monsoon one. The occurrence of uncertainties associated with rain-bearing cyclonic disturbances in the Bay of Bengal during the post-monsoon season may have caused the rainfall data distribution to become skewed.

On the other hand, the year-wise monsoon runoff and annual runoff series are found to be normally distributed, whereas all other seasonal series, namely pre-monsoon, post-monsoon, and winter ones do not follow normality. This deviation could be attributed to uncertainties in the occurrence of effective rainfall storm events during the non-monsoon periods. These findings suggested that the data series were normally distributed with a few exceptions that could be attributed to localized influence on the concerned parameter.

Table 4.2 Results of normality test for year-wise annual and seasonal rainfall (mm) and runoff (mm) total for the whole basin (1981–2018)

Variable	Time step	Kolmogorov–Smirnov[a]			Shapiro–Wilk		
		Statistic	*p*-Value	Remarks	Statistic	*p*-value	Remarks
Rainfall	Pre-monsoon	0.071	0.200*	Normal	0.983	0.821	Normal
	Monsoon	0.05	0.200*	Normal	0.986	0.895	Normal
	Post-monsoon	**0.159**	**0.017**	**NOT**	**0.815**	**0.000**	**NOT**
	Winter	0.12	0.187	Normal	0.953	0.113	Normal
	Annual	0.096	0.200*	Normal	0.975	0.532	Normal
Runoff	**Pre-monsoon**	**0.171**	**0.007**	**NOT**	**0.792**	**0.000**	**NOT**
	Monsoon	0.118	0.200*	Normal	0.969	0.368	Normal
	Post-monsoon	**0.199**	**0.001**	**NOT**	**0.849**	**0.000**	**NOT**
	Winter	**0.195**	**0.001**	**NOT**	**0.709**	**0.000**	**NOT**
	Annual	0.068	0.200*	Normal	0.984	0.851	Normal

*Lower bound of the true significance
[a]Lilliefors significance correction
Bold indicates statistical significance which refer to an observed pattern, trend, or change in a time series is unlikely to have occurred by chance alone; in this study, results with *p*-values less than 0.05 are considered statistically significant
Source Computed by the authors

Table 4.3 Results of normality test for year-wise seasonal and annual runoff (mm) for the whole basin (1973–2018)

Variable	Time step	Kolmogorov–Smirnov[a]			Shapiro–Wilk		
		Statistic	*p*-value	Remarks	Statistic	*p*-value	Remarks
Runoff	**Pre-monsoon**	**0.155**	**0.007**	**NOT**	**0.802**	**0.000**	**NOT**
	Monsoon	0.091	0.200*	Normal	0.970	0.289	Normal
	Post-monsoon	**0.208**	**0.000**	**NOT**	**0.836**	**0.000**	**NOT**
	Winter	**0.164**	**0.003**	**NOT**	**0.740**	**0.000**	**NOT**
	Annual	0.063	0.200*	Normal	0.982	0.685	Normal

*Lower bound of the true significance
[a]Lilliefors significance correction
Bold indicates statistical significance which refer to an observed pattern, trend, or change in a time series is unlikely to have occurred by chance alone; in this study, results with *p*-values less than 0.05 are considered statistically significant
Source Computed by the authors

In addition, the normality check was also carried out for runoff datasets for the duration of 1973–2018, and the results are found to be as same as what was observed for the 1981–2018 series (Table 4.3).

The normality tests were also carried out for seasonal and annual rainfall series for respective grid points during the period 1981–2018. Here, the tests were applied for rainfall series available from 12 grid points which are designated by grid point numbers 2, 3, 6, 7–8, 11–14, 18–19, and 23 (Fig. 4.1). Grid points 1, 4–5, 10, 15–17, 20–22, 24, and 25, on the other hand, were excluded from the analysis since they have no influence on the current river basin so far, as known through Thiessen Polygon analysis.

However, the results reveal that the post-monsoon series at all grid points are normally distributed, except at grid points 2 and 3 (Table 4.4). At these two grid points, both the post-monsoon and winter series do not follow normality. Again, inconsistency is observed between the result of the Kolmogorov–Smirnov test and that of the Shapiro–Wilk test for the winter rainfall series in grid points 2 and 3. Here, the result of the Shapiro–Wilk test was accepted since the length of the data series is below 50, for which the Shapiro–Wilk test is usually suggested.

These results suggested that the statistical properties of the data series vary across different seasons and grid points, and the post-monsoon season warrants further investigation to understand the effects of cyclonic disturbances that occurred in the Bay of Bengal on the non-normal distribution of the data.

(b) *Result of Autocorrelation Analysis*

The year-wise seasonal and annual total rainfall and runoff series were checked for lag1 autocorrelation using Pearson's product-moment correlation coefficient, and subsequently, the significance of the correlation coefficient value (r) was judged by using a t-test.

Table 4.4 Results of normality test of grid point specific year-wise seasonal and annual rainfall total (mm) (1981–2018)

Grid point	Time step	Kolmogorov–Smirnov[a]			Shapiro–Wilk		
		Statistic	*p*-value	Remarks	Statistic	*p*-value	Remarks
Grid point 2	Pre-monsoon	0.074	0.200*	Normal	0.972	0.444	Normal
	Monsoon	0.101	0.200*	Normal	0.971	0.428	Normal
	Post-monsoon	**0.157**	**0.019**	**NOT**	**0.818**	**0.000**	**NOT**
	Winter	0.136	0.074	Normal	**0.942**	**0.048**	**NOT**
	Annual	0.109	0.200*	Normal	0.964	0.259	Normal
Grid point 3	Pre-monsoon	0.074	0.200*	Normal	0.972	0.444	Normal
	Monsoon	0.101	0.200*	Normal	0.971	0.428	Normal
	Post-monsoon	**0.157**	**0.019**	**NOT**	**0.818**	**0.000**	**NOT**
	Winter	0.136	0.074	Normal	**0.942**	**0.048**	**NOT**
	Annual	0.109	0.200*	Normal	0.964	0.259	Normal
Grid point 6	Pre-monsoon	0.056	0.200*	Normal	0.979	0.67	Normal
	Monsoon	0.061	0.200*	Normal	0.982	0.781	Normal
	Post-monsoon	**0.16**	**0.015**	**NOT**	**0.805**	**0.000**	**NOT**
	Winter	0.109	0.200*	Normal	0.945	0.06	Normal
	Annual	0.099	0.200*	Normal	0.978	0.654	Normal
Grid point 7	Pre-monsoon	0.063	0.200*	Normal	0.982	0.771	Normal
	Monsoon	0.068	0.200*	Normal	0.981	0.764	Normal
	Post-monsoon	**0.156**	**0.02**	**NOT**	**0.803**	**0.000**	**NOT**
	Winter	0.124	0.147	Normal	0.951	0.093	Normal
	Annual	0.088	0.200*	Normal	0.977	0.613	Normal
Grid point 8	Pre-monsoon	0.063	0.200*	Normal	0.982	0.771	Normal
	Monsoon	0.068	0.200*	Normal	0.981	0.764	Normal
	Post-monsoon	**0.156**	**0.02**	**NOT**	**0.803**	**0.000**	**NOT**
	Winter	0.124	0.147	Normal	0.951	0.093	Normal
	Annual	0.088	0.200*	Normal	0.977	0.613	Normal
Grid point 11	Pre-monsoon	0.107	0.200*	Normal	0.961	0.199	Normal
	Monsoon	0.081	0.200*	Normal	0.977	0.598	Normal
	Post-monsoon	**0.171**	**0.007**	**NOT**	**0.814**	**0.000**	**NOT**
	Winter	0.115	0.200*	Normal	0.951	0.097	Normal
	Annual	0.088	0.200*	Normal	0.968	0.332	Normal

(continued)

Table 4.4 (continued)

Grid point	Time step	Kolmogorov–Smirnov[a]			Shapiro–Wilk		
		Statistic	p-value	Remarks	Statistic	p-value	Remarks
Grid point 12	Pre-monsoon	0.078	0.200*	Normal	0.976	0.581	Normal
	Monsoon	0.072	0.200*	Normal	0.981	0.765	Normal
	Post-monsoon	**0.15**	**0.03**	**NOT**	**0.815**	**0.000**	**NOT**
	Winter	0.114	0.200*	Normal	0.95	0.088	Normal
	Annual	0.08	0.200*	Normal	0.976	0.563	Normal
Grid point 13	Pre-monsoon	0.078	0.200*	Normal	0.976	0.581	Normal
	Monsoon	0.072	0.200*	Normal	0.981	0.765	Normal
	Post-monsoon	**0.15**	**0.03**	**NOT**	**0.815**	**0.000**	**NOT**
	Winter	0.114	0.200*	Normal	0.95	0.088	Normal
	Annual	0.08	0.200*	Normal	0.976	0.563	Normal
Grid point 14	Pre-monsoon	0.125	0.143	Normal	0.969	0.37	Normal
	Monsoon	0.11	0.200*	Normal	0.963	0.238	Normal
	Post-monsoon	0.132	0.095	Normal	**0.837**	**0.000**	**NOT**
	Winter	0.12	0.18	Normal	0.953	0.109	Normal
	Annual	0.059	0.200*	Normal	0.979	0.665	Normal
Grid point 18	Pre-monsoon	0.091	0.200*	Normal	0.974	0.511	Normal
	Monsoon	0.084	0.200*	Normal	0.973	0.483	Normal
	Post-monsoon	**0.156**	**0.021**	**NOT**	**0.835**	**0.000**	**NOT**
	Winter	0.102	0.200*	Normal	0.961	0.21	Normal
	Annual	0.074	0.200*	Normal	0.966	0.289	Normal
Grid point 19	Pre-monsoon	0.096	0.200*	Normal	0.961	0.209	Normal
	Monsoon	0.111	0.200*	Normal	0.963	0.237	Normal
	Post-monsoon	**0.151**	**0.028**	**NOT**	**0.852**	**0.000**	**NOT**
	Winter	0.087	0.200*	Normal	0.968	0.331	Normal
	Annual	0.07	0.200*	Normal	0.982	0.802	Normal
Grid point 23	Pre-monsoon	0.083	0.200*	Normal	0.972	0.435	Normal
	Monsoon	0.068	0.200*	Normal	0.971	0.426	Normal
	Post-monsoon	**0.15**	**0.031**	**NOT**	**0.858**	**0.000**	**NOT**
	Winter	0.074	0.200*	Normal	0.965	0.269	Normal
	Annual	0.117	0.200*	Normal	0.963	0.23	Normal

*Lower bound of the true significance

[a]Lilliefors significance correction

Bold indicates statistical significance which refer to an observed pattern, trend, or change in a time series is unlikely to have occurred by chance alone; in this study, results with p-values less than 0.05 are considered statistically significant

Note Stations 1, 4–5, 9, 10, 15–17, 20, 21, 22, and 24–25 were excluded from the analysis because they have no influential area inside the basin boundary as per Thiessen Polygon analysis

Source Computed by the authors based on rainfall data from NASA power data access, Langley Research Centre

Table 4.5 Results of autocorrelation test for year-wise seasonal and annual rainfall (mm) and runoff (mm) total for the whole basin (1981–2018)

Variable	Time step	r*	p-value	Remarks
Rainfall	Pre-monsoon	−0.121	0.476	Independent series
	Monsoon	−0.013	0.941	Independent series
	Post-monsoon	0.093	0.585	Independent series
	Winter	−0.068	0.689	Independent series
	Annual	0.013	0.939	Independent series
Runoff	Pre-monsoon	0.019	0.91	Independent series
	Monsoon	0.265	0.113	Independent series
	Post-monsoon	0.007	0.966	Independent series
	Winter	0.297	0.074	Independent series
	Annual	0.333	0.044	Autocorrelated series

*Pearson's product-moment correlation with lag 1
Bold indicates statistical significance which refer to an observed pattern, trend, or change in a time series is unlikely to have occurred by chance alone; in this study, results with p-values less than 0.05 are considered statistically significant
Source Computed by the authors

The result shows that all seasonal and annual rainfall series do not have autocorrelation, whereas, in the case of runoff, only the annual series shows the presence of significant autocorrelation (Table 4.5).

The absence of autocorrelation in rainfall variable for seasonal and annual series indicate the normal progression of whether phenomena may get influenced by local and regional factors. The distribution of rainfall in the present study area was highly variable across the grid points and was influenced by local topography and geography. The absence of autocorrelation in the seasonal runoff series and its presence in the annual series could be attributed to the human interferences only since the rainfall, which is the causal factor for runoff, is found to be free from autocorrelation in both seasonal and annual time step.

Autocorrelation was also assessed for the runoff datasets for the period 1973–2018, and the results show an exact match with the autocorrelation results of runoff for the period of 1981–2018 (Table 4.6).

Grid point-wise autocorrelation tests were also performed for year-wise seasonal and annual rainfall total for the duration 1981–2018. The results reveal that the data series for all the grid points are found to be independent at a 95% confidence level (Table 4.7). This suggests that the rainfall value of a particular time step can be treated as one having no influence on the data value of the previous or following time step.

Table 4.6 Results of autocorrelation test for year-wise seasonal and annual runoff total (mm) for the whole basin (1973–2018)

Variable	Time step	r*	p-value	Remarks
Runoff	Pre-monsoon	0.009	0.953	Independent series
	Monsoon	0.256	0.090	Independent series
	Post-monsoon	−0.022	0.888	Independent series
	Winter	0.284	0.059	Independent series
	Annual	0.275	0.048	Autocorrelated series

*Pearson's product-moment correlation with lag 1 series
Source Computed by the authors

4.3.2 Results of Step Change Analysis

(a) *The Basin Average Rainfall Series*

The Pettitt's test and the Sequential Mann–Kendall test were applied only to the post-monsoon series (1981–2018), which was observed to be non-normally distributed. In contrast, all other seasonal series and annual series were normally distributed, and therefore, the Buishand U test, Buishand Range test, and Standard Normal Homogeneity test were applied for them to detect any significant step changes. However, the results indicate that no significant change points were found in all seasonal and annual rainfall series which do not follow normality, and the same is found for normally distributed rainfall series too (Tables 4.8 and 4.9).

(b) *Grid Point-Wise Rainfall Series*

Pettitt's test and Sequential Mann–Kendall test were applied separately to the data series of each grid point that showed a non-normal distribution, while the Buishand U test, Buishand Range test, and Standard Normal Homogeneity test were used for normally distributed series for the duration of 1981–2018. The results show that none of the data series produced any significant step jump for any grid point (Tables 4.10 and 4.11). So, rainfall, either grid point specific or basin average series, shows stationarity over time which indicates that the climate change process, specifically the process of changing rainfall patterns that are observed at the global or continental scales, is absent in the concerned river basin.

(c) *The Runoff Series*

Pettitt test and Sequential Mann–Kendall Test were applied to the non-normal data, while the Buishand U test, Buishand Range test, and Standard Normal Homogeneity test were

Table 4.7 Results of autocorrelation test of station-specific year-wise seasonal and annual rainfall total (mm) (1981–2018)

Grid point	Time step	r^*	p-value	Remarks
Grid point 2	Pre-monsoon	0.123	0.467	Independent series
	Monsoon	−0.097	0.567	Independent series
	Post-monsoon	0.061	0.719	Independent series
	Winter	0.03	0.861	Independent series
	Annual	−0.027	0.874	Independent series
Grid point 3	Pre-monsoon	0.123	0.467	Independent series
	Monsoon	−0.097	0.567	Independent series
	Post-monsoon	0.061	0.719	Independent series
	Winter	0.03	0.861	Independent series
	Annual	−0.027	0.874	Independent series
Grid point 6	Pre-monsoon	−0.127	0.454	Independent series
	Monsoon	−0.018	0.914	Independent series
	Post-monsoon	0.095	0.576	Independent series
	Winter	−0.006	0.971	Independent series
	Annual	0.025	0.885	Independent series
Grid point 7	Pre-monsoon	−0.073	0.668	Independent series
	Monsoon	−0.036	0.834	Independent series
	Post-monsoon	0.095	0.575	Independent series
	Winter	−0.05	0.767	Independent series
	Annual	0.004	0.982	Independent series
Grid point 8	Pre-monsoon	−0.073	0.668	Independent series
	Monsoon	−0.036	0.834	Independent series
	Post-monsoon	0.095	0.575	Independent series
	Winter	−0.05	0.767	Independent series
	Annual	0.004	0.982	Independent series
Grid point 11	Pre-monsoon	−0.239	0.155	Independent series
	Monsoon	−0.021	0.903	Independent series
	Post-monsoon	0.101	0.544	Independent series
	Winter	−0.03	0.861	Independent series
	Annual	0.01	0.954	Independent series
Grid point 12	Pre-monsoon	−0.141	0.404	Independent series

(continued)

Table 4.7 (continued)

Grid point	Time step	r*	p-value	Remarks
	Monsoon	−0.006	0.973	Independent series
	Post-monsoon	0.103	0.545	Independent series
	Winter	−0.074	0.662	Independent series
	Annual	0.013	0.941	Independent series
Grid point 13	Pre-monsoon	−0.141	0.404	Independent series
	Monsoon	−0.006	0.973	Independent series
	Post-monsoon	0.103	0.545	Independent series
	Winter	−0.074	0.662	Independent series
	Annual	0.013	0.941	Independent series
Grid point 14	Pre-monsoon	0	1	Independent series
	Monsoon	0.042	0.806	Independent series
	Post-monsoon	0.076	0.655	Independent series
	Winter	−0.12	0.477	Independent series
	Annual	0.066	0.698	Independent series
Grid point 18	Pre-monsoon	−0.178	0.293	Independent series
	Monsoon	−0.008	0.961	Independent series
	Post-monsoon	0.08	0.639	Independent series
	Winter	−0.071	0.678	Independent series
	Annual	0.012	0.942	Independent series
Grid point 19	Pre-monsoon	0.019	0.911	Independent series
	Monsoon	0.065	0.702	Independent series
	Post-monsoon	0.057	0.736	Independent series
	Winter	−0.123	0.468	Independent series
	Annual	0.096	0.573	Independent series
Grid point 23	Pre-monsoon	−0.243	0.147	Independent series
	Monsoon	−0.049	0.774	Independent series
	Post-monsoon	0.04	0.816	Independent series
	Winter	−0.071	0.675	Independent series
	Annual	−0.013	0.94	Independent series

*Pearson's product-moment correlation with lag 1
Source Computed by the authors

Table 4.8 Results of step change analysis for year-wise seasonal and annual rainfall (mm) and runoff (mm) total for the entire basin for non-normal series (1981–2018)

Variable	Time step	Pettitt test				Sequential Mann–Kendall test	
		U* statistic	*p*-value	Change point	Remarks	Intersection	Change point (s)
Rainfall	Pre-monsoon	–	–	–	–	–	–
	Monsoon	–	–	–	–	–	–
	Post-monsoon	74	1	1992	Insignificant	TRUE	1982
	Winter	–	–	–	–	–	–
	Annual	–	–	–	–	–	–
Runoff	Pre-monsoon	108	0.577	2008	Insignificant	TRUE	2008, 2014, 2016
	Monsoon	–	–	–	–	–	–
	Post-monsoon	137	0.271	2007	Insignificant	TRUE	2014, 2016
	Winter	138	0.263	2008	Insignificant	TRUE	2015
	Annual	–	–	–	–	–	–

Source Computed by the authors

used for the normally distributed data for the duration 1981–2018. The results show that a significant change point was detected for the monsoon and annual runoff series. For other seasons namely pre-monsoon, post-monsoon, and winter, the series do not have significant change points (Tables 4.8 and 4.9).

The year 2001 was detected as a change point in the result of the Buishand *U* test and Buishand Range test, while 2008 was identified as a change point in the result of the Standard Normal Homogeneity test for both the series, i.e., monsoon and annual series.

Step change analysis was also conducted for the runoff datasets during 1973–2018. The results were like what was found for the period of 1981–2018. Only the monsoon and annual series had significant change points in 2001 and 2008. The Buishand *U* test and Buishand Range test identified the year 2001 as the change point, while the Standard Normal Homogeneity test detected the year 2008 as the change point for both monsoon and annual series (Tables 4.12 and 4.13).

Human intervention might be the only reason behind the presence of significant change points since rainfall did not exhibit any significant break in the time series of either seasonal or annual time steps. The LULC changes as well as the construction of canals, have resulted in a significant decrease in runoff. Two major projects, namely the Sankh-South Koel and South Koel-Subarnarekha interlinking projects, have played a crucial role

Table 4.9 Results of step change analysis for year-wise seasonal and annual rainfall (mm) and runoff (mm) total for the entire basin for normally distributed series (1981–2018)

Variable	Time step	Buishand *U* test			Buishand range test			Standard normal homogeneity test			Remarks
		U statistic	*p*-value	Change point	R statistic	*p*-value	Change point	T statistic	*p*-value	Change point	
Rainfall	Pre-monsoon	0.182	0.308	2005	1.32	0.174	2005	5.415	0.233	1981	NOT
	Monsoon	0.315	0.123	1993	1.145	0.375	1993	4.936	0.287	1993	NOT
	Post-monsoon	–	–	–	–	–	–	–	–	–	–
	Winter	0.212	0.245	1998	1.053	0.515	1998	5.763	0.197	1986	NOT
	Annual	0.341	0.098	1993	1.273	0.214	1993	5.71	0.199	1993	NOT
Runoff	Pre-monsoon	–	–	–	–	–	–	–	–	–	–
	Monsoon	**0.801**	**0.004**	**2001**	**1.691**	**0.015**	**2001**	**11.122**	**0.008**	**2008**	**Significant**
	Post-monsoon	–	–	–	–	–	–	–	–	–	–
	Winter	–	–	–	–	–	–	–	–	–	–
	Annual	**0.855**	**0.003**	**2001**	**1.747**	**0.009**	**2001**	**12.58**	**0.004**	**2008**	**Significant**

Bold indicates statistical significance which refer to an observed pattern, trend, or change in a time series is unlikely to have occurred by chance alone; in this study, results with *p*-values less than 0.05 are considered statistically significant
Source Computed by the authors

Table 4.10 Results of step change analysis of grid point specific year-wise seasonal and annual rainfall total (mm) for non-normal series (1981–2018)

Grid point	Time step	Pettitt's test				Sequential Mann–Kendall test	
		U* statistic	*p*-value	Change point	Remarks	Intersection	Change point (s)
2	Pre-monsoon	–	–	–	–	–	–
	Monsoon	–	–	–	–	–	–
	Post-monsoon	82	0.977	1992	Insignificant	TRUE	2006 and 2007
	Winter	128	0.349	1998	Insignificant	TRUE	1983, 2014 and 2015
	Annual	–	–	–	–	–	–
3	Pre-monsoon	–	–	–	–	–	–
	Monsoon	–	–	–	–	–	–
	Post-monsoon	82	0.977	1992	Insignificant	TRUE	2006 and 2007
	Winter	128	0.349	1998	Insignificant	TRUE	1983, 2014 and 2015
	Annual	–	–	–	–	–	–
6	Pre-monsoon	–	–	–	–	–	–
	Monsoon	–	–	–	–	–	–
	Post-monsoon	76	1	2004	Insignificant	TRUE	1982
	Winter	–	–	–	–	–	–
	Annual	–	–	–	–	–	–
7	Pre-monsoon	–	–	–	–	–	–
	Monsoon	–	–	–	–	–	–
	Post-monsoon	70	1	1992	Insignificant	TRUE	1982
	Winter	–	–	–	–	–	–
	Annual	–	–	–	–	–	–
8	Pre-monsoon	–	–	–	–	–	–
	Monsoon	–	–	–	–	–	–
	Post-monsoon	70	1	1992	Insignificant	TRUE	1982
	Winter	–	–	–	–	–	–

(continued)

Table 4.10 (continued)

Grid point	Time step	Pettitt's test				Sequential Mann–Kendall test	
		U* statistic	*p*-value	Change point	Remarks	Intersection	Change point (s)
	Annual	–	–	–	–	–	–
11	Pre-monsoon	–	–	–	–	–	–
	Monsoon	–	–	–	–	–	–
	Post-monsoon	65	1	1999	Insignificant	FALSE	–
	Winter	–	–	–	–	–	–
	Annual	–	–	–	–	–	–
12	Pre-monsoon	–	–	–	–	–	–
	Monsoon	–	–	–	–	–	–
	Post-monsoon	74	1	1992	Insignificant	TRUE	1982
	Winter	–	–	–	–	–	–
	Annual	–	–	–	–	–	–
13	Pre-monsoon	–	–	–	–	–	–
	Monsoon	–	–	–	–	–	–
	Post-monsoon	74	1	1992	Insignificant	TRUE	1982
	Winter	–	–	–	–	–	–
	Annual	–	–	–	–	–	–
14	Pre-monsoon	–	–	–	–	–	–
	Monsoon	–	–	–	–	–	–
	Post-monsoon	74	1	1992	Insignificant	TRUE	1982, 2015 and 2016
	Winter	–	–	–	–	–	–
	Annual	–	–	–	–	–	–
18	Pre-monsoon	–	–	–	–	–	–
	Monsoon	–	–	–	–	–	–
	Post-monsoon	72	1	1992	Insignificant	TRUE	1982
	Winter	–	–	–	–	–	–
	Annual	–	–	–	–	–	–

(continued)

Table 4.10 (continued)

Grid point	Time step	Pettitt's test				Sequential Mann–Kendall test	
		U* statistic	*p*-value	Change point	Remarks	Intersection	Change point (s)
19	Pre-monsoon	–	–	–	–	–	–
	Monsoon	–	–	–	–	–	–
	Post-monsoon	68	1	1984	Insignificant	TRUE	1982, 2015 and 2017
	Winter	–	–	–	–	–	–
	Annual	–	–	–	–	–	–
23	Pre-monsoon	–	–	–	–	–	–
	Monsoon	–	–	–	–	–	–
	Post-monsoon	87	0.893	1989	Insignificant	TRUE	1982, 2015 and 2017
	Winter	–	–	–	–	–	–
	Annual	–	–	–	–	–	–

Source Computed by the authors

in this regard. These projects were implemented to transfer water for various purposes like irrigation, electricity generation, and drinking water supply.

4.3.3 Results of Trend Analysis

(a) *The Basin Average Rainfall Series*

Since all the seasonal and annual rainfall series for the period 1981–2018 were found to be independent, the Mann–Kendall test was applied to capture the trend. The test results show that no significant trend was observed for any series, i.e., the data of seasonal or annual time step (Table 4.14). This absence of trend might be due to the fluctuating behavior of rainfall over time, in which the two opposite temporal responses might have made the series trendless.

To further investigate the absence of trends, the concerned trendless series were divided into two parts based on an eye-estimated breakpoint identified in the scatterplot, and trend analysis was reapplied to each segment separately. Only the pre-monsoon and winter series could be divided based on the eye-estimated breakpoint, while such division was not feasible for the other series. Therefore, those series were excluded from this analysis. The

Table 4.11 Results of step change analysis of grid point specific year-wise seasonal and annual rainfall total (mm) for normally distributed series (1981–2018)

Grid point	Time step	Buishand U test			Buishand range test			Standard normal homogeneity test			Remarks
		U statistic	*p*-value	Change point	R statistic	*p*-value	Change point	T statistic	*p*-value	Change point	
2	Pre-monsoon	0.299	0.130	1997	1.425	0.099	1997	5.185	0.262	1997	NOT
	Monsoon	0.171	0.332	1993	0.985	0.626	1993	2.927	0.666	1993	NOT
	Post-monsoon	–	–	–	–	–	–	–	–	–	–
	Winter	–	–	–	–	–	–	–	–	–	–
	Annual	0.212	0.252	1993	1.154	0.365	1993	4.138	0.407	1993	NOT
3	Pre-monsoon	0.299	0.130	1997	1.425	0.099	1997	5.185	0.262	1997	NOT
	Monsoon	0.171	0.332	1993	0.985	0.626	1993	2.927	0.666	1993	NOT
	Post-monsoon	–	–	–	–	–	–	–	–	–	–
	Winter	–	–	–	–	–	–	–	–	–	–
	Annual	0.212	0.252	1993	1.154	0.365	1993	4.138	0.407	1993	NOT
6	Pre-monsoon	0.042	0.938	1997	0.982	0.625	1997	3.834	0.475	1982	NOT
	Monsoon	0.319	0.115	1993	1.152	0.37	1993	4.721	0.323	1993	NOT
	Post-monsoon	–	–	–	–	–	–	–	–	–	–
	Winter	0.204	0.256	1998	1.048	0.523	1998	7.207	0.091	1982	NOT
	Annual	0.29	0.143	1993	1.228	0.27	1993	5.236	0.254	1992	NOT
7	Pre-monsoon	0.171	0.335	1997	1.267	0.221	1997	3.422	0.562	1982	NOT
	Monsoon	0.267	0.171	1993	1.146	0.378	1993	4.514	0.354	1993	NOT

(continued)

Table 4.11 (continued)

Grid point	Time step	Buishand U test			Buishand range test			Standard normal homogeneity test			Remarks
		U statistic	*p*-value	Change point	R statistic	*p*-value	Change point	T statistic	*p*-value	Change point	
	Post-monsoon	–	–	–	–	–	–	–	–	–	–
	Winter	0.233	0.217	1998	1.028	0.553	1998	7.396	0.084	1982	NOT
	Annual	0.293	0.14	1993	1.274	0.218	1993	5.409	0.233	1993	NOT
8	Pre-monsoon	0.171	0.335	1997	1.267	0.221	1997	3.422	0.562	1982	NOT
	Monsoon	0.267	0.171	1993	1.146	0.378	1993	4.514	0.354	1993	NOT
	Post-monsoon	–	–	–	–	–	–	–	–	–	–
	Winter	0.233	0.217	1998	1.028	0.553	1998	7.396	0.084	1982	NOT
	Annual	0.293	0.14	1993	1.274	0.218	1993	5.409	0.233	1993	NOT
11	Pre-monsoon	0.04	0.945	1982	0.934	0.706	1982	5.741	0.199	1982	NOT
	Monsoon	0.339	0.106	1993	1.161	0.351	1993	5.439	0.23	1992	NOT
	Post-monsoon	–	–	–	–	–	–	–	–	–	–
	Winter	0.195	0.28	1998	1.007	0.59	1998	5.57	0.213	1986	NOT
	Annual	0.295	0.137	1993	1.218	0.275	1993	5.7	0.201	1992	NOT
12	Pre-monsoon	0.166	0.353	2005	1.299	0.191	2005	5.574	0.217	1981	NOT
	Monsoon	0.323	0.117	1993	1.154	0.362	1993	5.131	0.261	1989	NOT
	Post-monsoon	–	–	–	–	–	–	–	–	–	–

(continued)

Table 4.11 (continued)

Grid point	Time step	Buishand U test			Buishand range test			Standard normal homogeneity test			Remarks
		U statistic	*p*-value	Change point	R statistic	*p*-value	Change point	T statistic	*p*-value	Change point	
	Winter	0.206	0.262	1998	1.065	0.498	1998	5.589	0.211	1986	NOT
	Annual	0.343	0.101	1993	1.27	0.219	1993	5.777	0.194	1993	NOT
13	Pre-monsoon	0.166	0.353	2005	1.299	0.191	2005	5.574	0.217	1981	NOT
	Monsoon	0.323	0.117	1993	1.154	0.362	1993	5.131	0.261	1989	NOT
	Post-monsoon	–	–	–	–	–	–	–	–	–	–
	Winter	0.206	0.262	1998	1.065	0.498	1998	5.589	0.211	1986	NOT
	Annual	0.343	0.101	1993	1.27	0.219	1993	5.777	0.194	1993	NOT
14	Pre-monsoon	0.361	0.091	2005	1.456	0.083	2005	7.353	0.082	2014	NOT
	Monsoon	0.375	0.082	2005	1.093	0.454	2005	4.664	0.33	2005	NOT
	Post-monsoon	–	–	–	–	–	–	–	–	–	–
	Winter	0.226	0.226	1998	1.079	0.473	1998	5.856	0.189	1986	NOT
	Annual	0.471	0.054	1993	1.254	0.235	1993	5.6	0.213	1993	NOT
18	Pre-monsoon	0.158	0.374	2005	1.252	0.242	2005	7.423	0.082	1981	NOT
	Monsoon	0.335	0.104	1993	1.088	0.456	1993	5.186	0.261	1989	NOT
	Post-monsoon	–	–	–	–	–	–	–	–	–	–
	Winter	0.17	0.343	1998	1.064	0.492	1998	5.152	0.257	1986	NOT

(continued)

Table 4.11 (continued)

Grid point	Time step	Buishand U test			Buishand range test			Standard normal homogeneity test			Remarks
		U statistic	*p*-value	Change point	R statistic	*p*-value	Change point	T statistic	*p*-value	Change point	
	Annual	0.35	0.095	1993	1.21	0.287	1993	5.728	0.197	1992	NOT
19	Pre-monsoon	0.346	0.101	2012	1.504	0.058	2012	8.327	0.051	2014	NOT
	Monsoon	0.411	0.066	2005	1.105	0.431	2005	5.103	0.264	2005	NOT
	Post-monsoon	–	–	–	–	–	–	–	–	–	–
	Winter	0.183	0.308	1998	1.039	0.539	1998	5.618	0.21	1986	NOT
	Annual	0.528	0.059	1993	1.234	0.258	1993	5.592	0.215	1993	NOT
23	Pre-monsoon	0.113	0.528	2012	1.211	0.291	2012	8.392	0.057	1981	NOT
	Monsoon	0.283	0.147	1992	0.983	0.627	1992	4.388	0.376	1992	NOT
	Post-monsoon	–	–	–	–	–	–	–	–	–	–
	Winter	0.125	0.484	1998	1.009	0.593	1998	4.375	0.373	1986	NOT
	Annual	0.301	0.131	1992	1.117	0.415	1992	5.099	0.261	1992	NOT

Source Computed by the authors

Table 4.12 Results of step change analysis for non-normal series of runoff (1973–2018)

Time step	Pettitt test				Sequential Mann–Kendall test	
	U* statistic	*p*-value	Change point	Remarks	Intersection	Change point (s)
Pre-monsoon	122	0.815	2008	Insignificant	TRUE	2009, 2014 and 2016
Monsoon	–	–	–	–	–	–
Post-monsoon	163	0.403	2007	Insignificant	TRUE	1974, 1985, 2014 and 2016
Winter	174	0.322	2008	Insignificant	TRUE	1974, 1977 and 2015
Annual	–	–	–	–	–	–

Source Computed by the authors

results suggested that the rainfall during pre-monsoon significantly increased at the rate of 1.692 mm/year since 1991 at a 95% confidence level whereas the same variable during before 1991 was found to be trendless (Table 4.15 and Fig. 4.2). The winter rainfall, on the other hand, shows a significant decreasing trend (at the rate of 5.8 mm/year) before 1994, and from 1994 onward, the series is found to be trendless.

(b) *Grid Point-Wise Rainfall Series*

The Mann–Kendall test was applied to capture the trend in rainfall time series for respective grid points since there was no autocorrelation in all the series. The results reveal that, in general, there was no trend in the data series for all variables, except for a few grid points series (Table 4.16). The annual rainfall at grid point 14 and grid point 19 exhibited a significant increasing trend at a rate of 8.460 and 9.523 mm/year, respectively. Additionally, the rainfall data for grid point 14 and grid point 23 during the pre-monsoon and monsoon seasons respectively also demonstrated a significant increasing trend. The rate of increase was 1.290 and 5.004 mm/year, respectively.

(c) *The Runoff Series*

The trend analysis of the pre-monsoon, monsoon, post-monsoon, and winter time series for the period 1981–2018 was carried out using the Mann–Kendall test since the concerned series do not have autocorrelation (see Table 4.5). The annual runoff series, on the other hand, was analysed by using the Modified Mann–Kendall test as the concerned series has autocorrelation. The results show significant decreasing trends in monsoon and annual runoff series at the rate of 5.290 and 5.929 mm/year (Table 4.14 and Fig. 4.3).

Table 4.13 Results of step change analysis for normal series of runoff (1973–2018)

Time step	Buishand U test			Buishand range test			Standard normal homogeneity test			Remarks
	U statistic	*p*-value	Change point	R statistic	*p*-value	Change point	T statistic	p-value	Change point	
Pre-monsoon	–	–	–	–	–	–	–	–	–	–
Monsoon	**0.785**	**0.006**	**2001**	**1.702**	**0.018**	**2001**	**11.627**	**0.009**	**2008**	**Significant**
Post-monsoon	–	–	–	–	–	–	–	–	–	–
Winter	–	–	–	–	–	–	–	–	–	–
Annual	**0.802**	**0.005**	**2001**	**1.627**	**0.031**	**2001**	**12.988**	**0.004**	**2008**	**Significant**

Bold indicates statistical significance which refer to an observed pattern, trend, or change in a time series is unlikely to have occurred by chance alone; in this study, results with *p*-values less than 0.05 are considered statistically significant

Source Computed by the authors

Table 4.14 Results of trend analysis of rainfall and runoff data

Variable	Time step	Mann–Kendall test		Modified Mann–Kendall test		Sen's slope	Remarks
		z statistic	*p*-value	Zc statistic	*p*-value		
Rainfall	Pre-monsoon	1.584	0.113	–	–	0.864	NOT
	Monsoon	1.685	0.092	–	–	5.8	NOT
	Post-monsoon	0.402	0.688	–	–	0.263	NOT
	Winter	−1.182	0.237	–	–	−0.584	NOT
	Annual	1.685	0.092	–	–	6.069	NOT
Runoff	Pre-monsoon	−1.106	0.267	–	–	−0.042	NOT
	Monsoon	**−2.439**	**0.015**	**–**	**–**	**−5.29**	**Significant**
	Post-monsoon	−0.754	0.451	–	–	−0.194	NOT
	Winter	−1.207	0.228	–	–	−0.097	NOT
	Annual	**–**	**–**	**−2.364**	**0.018**	**−5.929**	**Significant**

Bold indicates statistical significance which refer to an observed pattern, trend, or change in a time series is unlikely to have occurred by chance alone; in this study, results with *p*-values less than 0.05 are considered statistically significant

Source Computed by the authors

Table 4.15 Results of trend analysis of rainfall (after series partitioning)

Variable	Time step	Period	Mann–Kendall test		Sen's slope	Remarks
			z statistic	*p*-value		
Rainfall	Pre-monsoon	1981–1990	−0.894	0.371	−4.594	NOT
		1991–2018	**2.351**	**0.019**	**1.692**	**Significant**
	Monsoon	–	–	–	–	–
	Post-monsoon	–	–	–	–	–
	Winter	**1981–1993**	**−2.745**	**0.006**	**−5.8**	**Significant**
		1994–2018	−0.49	0.623	−0.556	NOT
	Annual	–	–	–	–	–

Bold indicates statistical significance which refer to an observed pattern, trend, or change in a time series is unlikely to have occurred by chance alone; in this study, results with *p*-values less than 0.05 are considered statistically significant

Source Computed by the authors

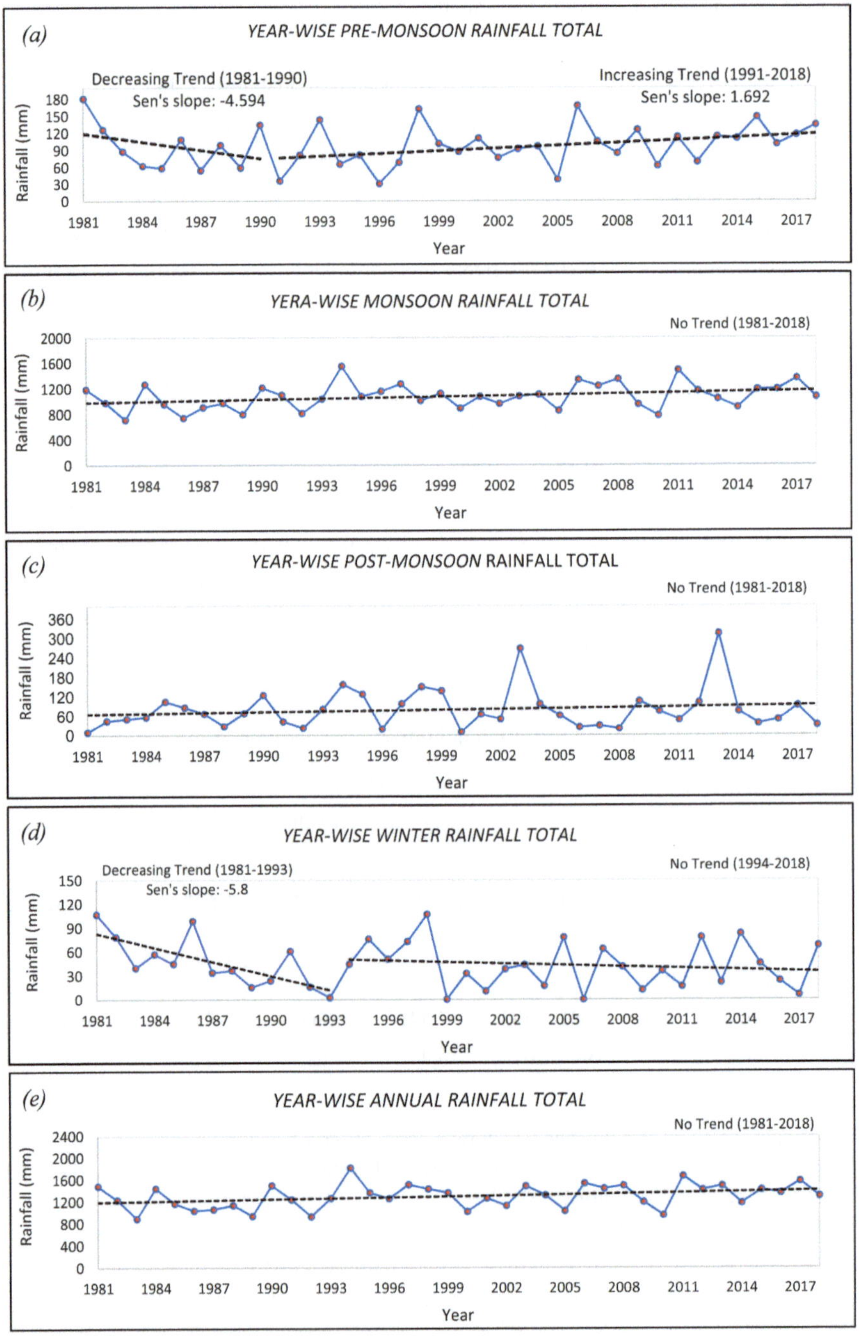

Fig. 4.2 Trend in rainfall for **a** pre-monsoon, **b** monsoon, **c** post-monsoon, **d** winter, and **e** annual series (1981–2018). *Source* Computed based on NASA Power Data Access (2020)

Table 4.16 Results of grid point-wise trend analysis in rainfall series

Grid point	Time step	Mann–Kendall test		Modified Mann–Kendall test		Sen's slope	Remarks
		z statistic	*p*-value	Zc statistic	*p*-value		
2	Pre-monsoon	1.308	0.191	–	–	0.528	NOT
	Monsoon	1.132	0.258	–	–	4.314	NOT
	Post-monsoon	0	1	–	–	−0.001	NOT
	Winter	1.484	0.138	–	–	−0.627	NOT
	Annual	1.358	0.175	–	–	4.588	NOT
3	Pre-monsoon	1.308	0.191	–	–	0.528	NOT
	Monsoon	1.132	0.258	–	–	4.314	NOT
	Post-monsoon	0	1	–	–	−0.001	NOT
	Winter	1.484	0.138	–	–	−0.627	NOT
	Annual	1.358	0.175	–	–	4.588	NOT
6	Pre-monsoon	0.201	0.841	–	–	0.088	NOT
	Monsoon	1.685	0.092	–	–	6.07	NOT
	Post-monsoon	0.025	0.98	–	–	0.003	NOT
	Winter	−1.182	0.237	–	–	−0.5823	NOT
	Annual	1.408	0.159	–	–	5.738	NOT
7	Pre-monsoon	1.056	0.291	–	–	0.6	NOT
	Monsoon	1.66	0.097	–	–	4.938	NOT
	Post-monsoon	0.201	0.841	–	–	0.169	NOT
	Winter	−1.31	0.191	–	–	−0.593	NOT
	Annual	1.56	0.119	–	–	5.181	NOT
8	Pre-monsoon	1.056	0.291	–	–	0.6	NOT
	Monsoon	1.66	0.097	–	–	4.938	NOT
	Post-monsoon	0.201	0.841	–	–	0.169	NOT
	Winter	−1.31	0.191	–	–	−0.593	NOT
	Annual	1.56	0.119	–	–	5.181	NOT
11	Pre-monsoon	0.4526	0.651	–	–	0.191	NOT
	Monsoon	1.76	0.078	–	–	5.836	NOT
	Post-monsoon	0	1	–	–	−0.007	NOT
	Winter	−1.157	0.247	–	–	−0.544	NOT
	Annual	1.76	0.078	–	–	5.55	NOT

(continued)

Table 4.16 (continued)

Grid point	Time step	Mann–Kendall test		Modified Mann–Kendall test		Sen's slope	Remarks
		z statistic	*p*-value	Zc statistic	*p*-value		
12	Pre-monsoon	1.458	0.145	–	–	0.839	NOT
	Monsoon	1.634	0.102	–	–	5.526	NOT
	Post-monsoon	0.477	0.633	–	–	0.308	NOT
	Winter	−1.144	0.253	–	–	−0.55	NOT
	Annual	1.71	0.087	–	–	5.982	NOT
13	Pre-monsoon	1.458	0.145	–	–	0.839	NOT
	Monsoon	1.634	0.102	–	–	5.526	NOT
	Post-monsoon	0.477	0.633	–	–	0.308	NOT
	Winter	−1.144	0.253	–	–	−0.55	NOT
	Annual	1.71	0.087	–	–	5.982	NOT
14	**Pre-monsoon**	**1.961**	**0.05**	**–**	**–**	**1.29**	**Significant**
	Monsoon	1.684	0.092	–	–	5.88	NOT
	Post-monsoon	0.754	0.451	–	–	0.651	NOT
	Winter	−1.357	0.175	–	–	−0.651	NOT
	Annual	**1.961**	**0.05**	**–**	**–**	**8.46**	**Significant**
18	Pre-monsoon	1.282	0.2	–	–	0.894	NOT
	Monsoon	1.785	0.074	–	–	5.616	NOT
	Post-monsoon	0.553	0.58	–	–	0.281	NOT
	Winter	−1.031	0.303	–	–	−0.563	NOT
	Annual	1.861	0.063	–	–	6.241	NOT
19	Pre-monsoon	1.71	0.087	–	–	1.432	NOT
	Monsoon	1.836	0.066	–	–	7.059	NOT
	Post-monsoon	0.654	0.513	–	–	0.613	NOT
	Winter	−1.157	0.247	–	–	−0.588	NOT
	Annual	**2.137**	**0.033**	**–**	**–**	**9.523**	**Significant**
23	Pre-monsoon	1.056	0.291	–	–	0.789	NOT
	Monsoon	**1.961**	**0.05**	**–**	**–**	**5.004**	**Significant**
	Post-monsoon	0.502	0.615	–	–	0.384	NOT
	Winter	−0.83	0.407	–	–	−0.349	NOT
	Annual	1.861	0.063	–	–	4.911	NOT

Bold indicates statistical significance which refer to an observed pattern, trend, or change in a time series is unlikely to have occurred by chance alone; in this study, results with *p*-values less than 0.05 are considered statistically significant

Source Computed by the authors

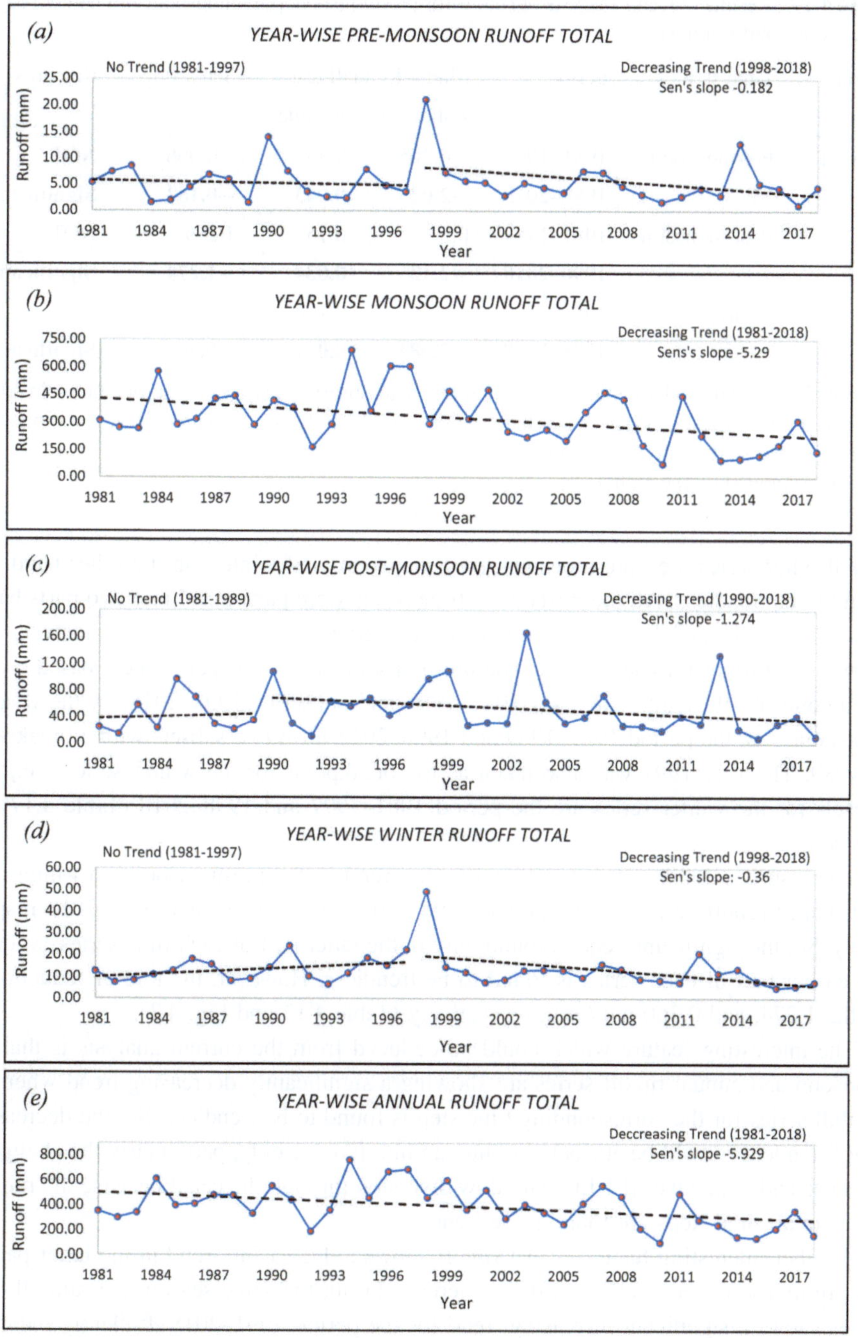

Fig. 4.3 Trend in the runoff for **a** pre-monsoon, **b** monsoon, **c** post-monsoon, **d** winter, and **e** annual series (1981–2018). *Source* Computed based on CWC discharge data (2020)

Table 4.17 Results of trend analysis of runoff for pre-monsoon, post-monsoon, and winter seasons (after series partitioning)

Variable	Time step	Period	Mann–Kendall test		Sen's slope	Remarks
			z statistic	*p*-value		
Runoff	Pre-monsoon	1981–1997	−0.205	0.837	−0.066	NOT
		1998–2018	**−2.023**	**0.043**	**−0.182**	**Significant**
	Post-monsoon	1981–1989	0.313	0.755	1.279	NOT
		1990–2018	**−2.12**	**0.034**	**−1.274**	**Significant**
	Winter	1981–1997	1.607	0.108	0.469	NOT
		1998–2018	**−2.084**	**0.037**	**−0.36**	**Significant**

Bold indicates statistical significance which refer to an observed pattern, trend, or change in a time series is unlikely to have occurred by chance alone; in this study, results with *p*-values less than 0.05 are considered statistically significant
Source Computed by the authors

All other series, i.e., pre-monsoon, post-monsoon, and winter, on the other hand, are found to be trendless. All these trendless time series were partitioned into two parts based on the eye-estimated breakpoint found in the scatter plot, and then trend analysis was carried out further for each subset. The pre-monsoon series was partitioned, based on the breakpoint at 1997, into subseries 1981–1997 and subseries 1998–2018. In the case of post-monsoon, the period 1981–1989 and 1990–2018 form two subsets with a breakpoint at 1989. The year 1997 was also taken as the breakpoint for the winter series, i.e., two subsets for the winter series are the period 1981–1997 and 1998–2018 (Table 4.17 and Fig. 4.3).

The results exhibit a significant decreasing trend in the pre-monsoon, post-monsoon, and winter runoff series for the period 1998–2018, 1990–2018, and 1998–2018 respectively. So, the significant trend is found only in the latter part of each time series, whereas the earlier part of each series is found to be trendless. However, the rate of reduction is 0.182, 1.274, and 0.360 mm/year, respectively (Table 4.17 and Fig. 4.3).

The interesting feature which could be deduced from the current analysis is that the monsoon and annual runoff series are showing a significantly decreasing trend when the rainfall series for the corresponding time step is found to be trendless. So, the decreasing runoff tendency could be attributed to human interference only, particularly the change in land use and land cover (LULC) and flow diversion through the canal because the rainfall series in all time steps are found to be trendless.

Another interesting feature is that runoff shows a decreasing trend in the latter part of the pre-monsoon series (1998–2018), whereas, during the same season, the rainfall time series shows a significant increasing trend for the period 1991–2018. It clearly indicates that this rainfall increase has no influence on runoff during the pre-monsoon season. Most of the rainwater goes out of the basin in the form of abstraction loss without producing runoff.

The latter part of the runoff series during post-monsoon (1990–2018) and winter (1998–2018) shows a decreasing trend, but the rainfall series for the corresponding seasons are showing trendless. Moreover, the winter rainfall series shows decreasing tendency in its earlier part (1981–1993). All these clearly indicate that the decreasing tendency in runoff has no connection with rainfall since these seasons are the dry seasons during which stream flow is supplied by groundwater (base flow) only. Hence, the decrease in runoff during winter and post-monsoon might be due to groundwater depletion that, on the other hand, might be due to groundwater withdrawal for its use in different sectors.

In addition, trend analysis was also performed for the runoff series for the period 1973–2018. The results are found as the same as obtained for the period of 1981–2018, i.e., the monsoon and annual series are showing a significant decreasing trend. The monsoon runoff is decreasing at the rate of 4.731 mm/year, while the rate of decrease in the annual series is 5.126 mm/year (Table 4.18).

In addition, trend analysis was also performed for the runoff series for the period 1973–2018. The results are found as the same as obtained for the period of 1981–2018, i.e., the monsoon and annual series are showing a significant decreasing trend. The monsoon runoff is decreasing at the rate of 4.731 mm/year, while the rate of decrease in the annual series is 5.126 mm/year (Table 4.18).

Another interesting feature that can be deduced from the current analysis is that the monsoon and annual runoff time series show significant step changes as well as trends. But based on the change point, if the series is partitioned into two subsets and trend analysis is carried out again for each subset, no trend is found at all for each series (Table 4.19).

Table 4.18 Results of trend analysis of runoff (1973–2018)

Variable	Time step	Mann–Kendall test		Modified Mann–Kendall test		Sen's slope	Remarks
		z statistic	*p*-value	Zc statistic	*p*-value		
Runoff	Pre-monsoon	−0.568	0.570	–	–	−0.015	NOT
	Monsoon	**−2.575**	**0.010**	–	–	**−4.731**	**Significant**
	Post-monsoon	−0.644	0.520	–	–	−0.142	NOT
	Winter	−1.136	0.256	–	–	−0.066	NOT
	Annual	–	–	**−2.632**	**0.008**	**−5.126**	**Significant**

Bold indicates statistical significance which refer to an observed pattern, trend, or change in a time series is unlikely to have occurred by chance alone; in this study, results with *p*-values less than 0.05 are considered statistically significant

Source Computed by the authors

Table 4.19 Results of trend analysis after series partitioning for runoff data

Time-step	Mann–Kendall test (monsoon series)				Mann–Kendall test (annual series)			
	z statistic	*p*-value	Sens's slope	Remarks	Zc statistic	*p*-value	Sens's slope	Remarks
1981–2001	1.661	0.097	5.16	Insignificant	1.842	0.065	8.767	NOT
2002–2018	−1.195	0.232	−7.833	Insignificant	−1.689	0.091	−11.932	NOT
1981–2008	0.059	0.953	0.279	Insignificant	0.533	0.594	1.821	NOT
2009–2018	0.358	0.721	7.21	Insignificant	−0.17	0.858	−5.744	NOT

Source Computed by the authors

The year 2001 was detected as a change point as per the results of the Buishand U test and Buishand Range test for both the series, i.e., monsoon and annual series. Based on the change point, two subsets are formed, i.e., 1981–2001 and 2002–2018. The year 2008, on the other hand, was identified as a change point in the result of the Standard Normal Homogeneity test for the same series, i.e., monsoon and annual series, and the year 2008 partitioned the concerned series into two subsets, i.e., the period 1981–2008 and 2009–2018. The Mann–Kendall test was used for capturing trends in each subset because there is no autocorrelation in the concerned time series. The results show that the runoff data series of each subset is found to be trendless (Table 4.19 and Fig. 4.4).

4.4 The Overall Discussion

The results of step change analysis of basin average rainfall show that there are no significant change points in seasonal and annual rainfall series. Moreover, the results of the grid point-wise rainfall series show that none of the data series produced any significant step jump for any grid point. So, rainfall, either grid point specific or basin average series, shows stationarity over time which indicates that the climate change process, specifically the process of changing rainfall patterns that are observed at the global or continental scale, is absent in the concerned river basin.

The results of the step change analysis of the runoff series for the period of 1981–2018 show that a significant change point was detected for the monsoon and annual runoff series. The monsoon and annual series both have significant change points in 2001 and 2008. For other seasons namely pre-monsoon, post-monsoon, and winter, the series does not have significant change points. Step change analysis of runoff was also conducted for the runoff datasets during 1973–2018. The results were similar to what was found for the period of 1981–2018. Human intervention might be the only reason behind the presence of significant change points since rainfall did not exhibit any significant break in the time series of either seasonal or annual time steps. The LULC changes as well as the construction of canals, have resulted in a significant decrease in runoff.

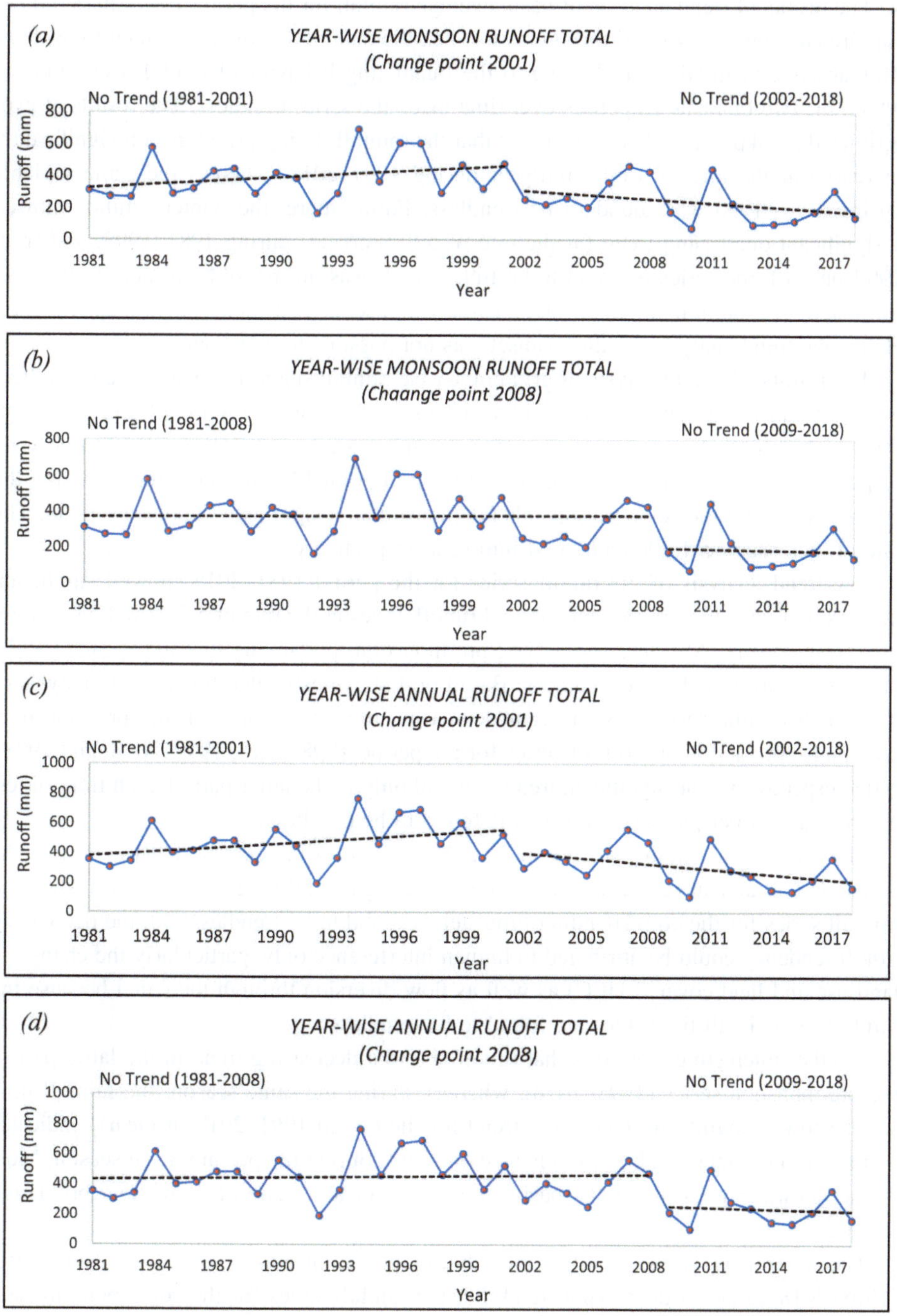

Fig. 4.4 Results of trend analysis after partitioning the runoff series based on **a** change point 2001, **b** change point 2008 in the monsoon series, **c** change point 2001, and **d** change point 2008 in the annual series. *Source* Computed based on CWC discharge data (2020)

The results of trend analysis of basin average rainfall for the period 1981–2018 did not capture any significant trend for any series, i.e., the data of seasonal or annual time steps. This absence of trend might be due to the fluctuating behavior of rainfall over time, in which the two opposite responses over time made the series trendless. The results of eye-estimated breakpoint analysis suggested that the rainfall during pre-monsoon significantly increased at the rate of 1.692 mm/year during 1991–2018, whereas the same variable during 1981–1990 was found to be trendless. Furthermore, the winter rainfall showed a significant decreasing trend (at the rate of 5.8 mm/year) during 1981–1993, and from 1994 onward, the series is found to be trendless. It was not possible to identify the eye-estimated breakpoint in all other series, namely monsoon, post-monsoon, and annual ones. Hence the trend analysis for their subset was not possible to carry out.

The results of trend analysis of grid point-wise rainfall suggested that, in general, there was no trend in the data series for all variables, except for grid point 14 and grid point 19, which exhibited a significant increasing trend at a rate of 8.460 and 9.523 mm/year, respectively. Additionally, the rainfall data for grid point 14 and grid point 23 during the pre-monsoon and monsoon seasons also demonstrated a significant increasing trend. The rate of increase was 1.290 and 5.004 mm/year, respectively.

The trend analysis of the runoff series for the period 1981–2018 showed significant decreasing trends in monsoon and annual runoff series at the rate of 5.290 and 5.929 mm/year, respectively. All other series, i.e., pre-monsoon, post-monsoon, and winter, on the other hand, are found to be trendless. Based on the eye-estimated breakpoint analysis of the trendless time series, a significant decreasing trend was found in the pre-monsoon, post-monsoon, and winter runoff series for the period 1998–2018, 1990–2018, and 1998–2018 respectively. The significant trend is found only in the latter part of each time series, whereas the earlier part of each series is found to be trendless.

The interesting feature which could be deduced from the current analysis is that the monsoon and annual runoff series are showing a significantly decreasing trend when the rainfall series for the corresponding time step is found to be trendless. So, the decreasing runoff tendency could be attributed to human interference only, particularly the change in land use and land cover (LULC) as well as flow diversion through the canal because the rainfall series in all time steps are found to be trendless.

Another interesting feature is that runoff shows a decreasing trend in the latter part of the pre-monsoon series (1998–2018), whereas, during the same season, the rainfall time series shows a significant increasing trend for the period 1991–2018. It clearly indicates that this rainfall increase has no influence on runoff during the pre-monsoon season. Most of the rainwater goes out of the basin in the form of abstraction loss without producing runoff.

The latter part of the runoff series during post-monsoon (1990–2018) and winter (1998–2018) shows a decreasing trend, but the rainfall series for the corresponding seasons are showing trendless. Moreover, the winter rainfall series shows decreasing tendency in its earlier part (1981–1993). All these clearly indicate that the decreasing tendency in runoff has no connection with rainfall since these seasons are the dry seasons during which stream flow is supplied by groundwater (base flow) only. Hence, the runoff decrease during winter and post-monsoon might be due to groundwater depletion that, on the other hand, might be due to groundwater withdrawal for its use in different sectors.

In addition, trend analysis was also performed for the runoff series for the period 1973–2018, and the results suggested that the monsoon and annual series are showing a significant decreasing trend. The monsoon runoff is decreasing at the rate of 4.731 mm/year, while the rate of decrease in the annual series is 5.126 mm/year.

Another interesting feature that can be deduced from the current analysis is that the monsoon and annual runoff time series show significant step changes as well as trends. But based on the change point, if the series is partitioned into two subsets and trend analysis is carried out again for each subset, no trend is found at all for each series.

The scenario that was deduced from the current analysis was not consistent with the global and hemispheric scale, but a bit of similarity was observed with the outcomes of the other works done in the Indian context. Guhathakurta and Rajeevan (2008) found that pre-monsoon rainfall which was caused by convective activity, decreased in central India, while annual rainfall decreased in Chhattisgarh, Jharkhand, and Kerala. Winter rainfall increased for some regions, including Himachal Pradesh and Tripura, but decreased significantly in east Uttar Pradesh, Bihar, and east Madhya Pradesh due to Western disturbances. Meanwhile, Kumar et al. (2010) reported no significant trends in annual, seasonal, or monthly rainfall for India as a whole but reported increases in pre-monsoon, post-monsoon, and winter rainfall and decreases in annual and monsoon rainfall in certain areas. Additionally, there was a decrease in rainfall in June, July, and September but an increase in August all across the country. These findings matched a bit with the current findings, although the time periods were different. However, Raha et al. (2014), Bisai (2016), and Das and Bhattacharya (2018) reported that the rainfall in different parts of India, particularly in the northern part of West Bengal, was slightly decreasing during monsoon and post-monsoon although the total amount of rain remains almost unchanged. Kulkarni et al. (2020) found a negative trend in all-India annual and pre-monsoon rainfall during 1951–2015, mainly over the Indo-Gangetic Plains and Western Ghats. Meanwhile, Gogoi and Rao (2022) found an increasing trend in annual, pre-monsoon, and monsoon rainfall over Assam from 1981 to 2017, while post-monsoon and winter seasons had no significant trends.

In Jharkhand, Sharma and Singh (2017) reported a significant decrease in annual, monsoon, and winter rainfall from 1901 to 2002, which could impact agricultural activities. However, Praveen et al. (2020) found that pre-monsoon and post-monsoon rainfall decreased while monsoon, winter, and annual rainfall increased during 1901–2015 in the same state. According to the report published by IMD (2020), there was a significant decrease in southwest monsoon rainfall over a 30-year period from 1989 to 2018 in states such as Uttar Pradesh, Bihar, West Bengal, Meghalaya, and Nagaland. Annual rainfall also decreased in these states, as well as in Arunachal Pradesh and Himachal Pradesh.

However, there were no significant changes observed in southwest monsoon rainfall in the other states during the same period. These results were identical to the findings of this study.

The findings of step change and trend analysis of runoff and its relation with the rainfall were not simple rather, it is complex interactions between climate variables, human activities, and other geophysical conditions. Several researchers reported that the runoff was changed in different parts of the world, including India. Bouwer et al. (2006) stated that runoff was decreased in the Krishna River basin because of largely due to the impact of human activities rather than rainfall change. Similar kind of results were found by Sharma et al. (2019). Verma and Dash (2018) reported exactly the opposite situation found in the NW Himalayan region. Swain and Patra (2019) stated that rainfall played a significant role in determining the runoff patterns over the catchments of eastern and southern India. The findings of runoff change in this study were not associated with the findings of Swain and Patra (2019). However, the findings of this study were best matched with that area where rainfall variability was less and human activities were more prominent. It can be stated that if the duration of the time series used in different studies is the same, it is possible to have similar results.

4.5 Findings in Brief

The results obtained from the analysis of temporal variability of seasonal and annual rainfall and runoff series of the concerned river basin are summarized below:

(i) The rainfall series, either basin average or grid point specific ones, in either seasonal and annual time steps do not show any significant change point for the period of 1981–2018.

(ii) Runoff, on the other hand, show significant step change in 2001 as per the result of the Buishand *U* and Buishand Range test, while as per the result of the Standard Normal Homogeneity test, the year 2008 was detected as a significant change point in monsoon and annual runoff series for the period of 1981–2018 and the same result is also observed for the period of 1973–2018.

(iii) So far as the rainfall trend is concerned, the basin average seasonal or annual rainfall series does not show any significant trend for the period of 1981–2018. The trend was also not observed in grid-specific seasonal and annual rainfall time series except for the series from grid point 14, grid point 19, and grid point 23. The pre-monsoon and annual series from grid point 14 show a significant increasing trend, whereas the annual series from grid point 19 and the monsoon series from grid point 23 show a significant increasing trend.

(iv) Runoff, on the other hand, shows significant decreasing trends only in monsoon (at the rate of 5.29 mm/year) and annual series (at the rate of 5.929 mm/year) for the period of 1981–2018, whereas other seasonal series are found trendless. Similar results are also observed for the period 1973–2018. Here, runoff during monsoon is decreasing at the rate of 4.731 mm/year while the rate of decrease in the annual series is 5.126 mm/year.

4.6 Concluding Remarks

Step change and trend analysis identified no significant change points and trends in the rainfall series, respectively. Unlike rainfall, significant change points and trends were found in the runoff series. The annual runoff has been decreasing day by day since 2002 onwards. So, this decrease in runoff is assumed to be largely due to other factors like human intervention. However, this runoff response can be properly substantiated in the context of human interference which was assessed in Chap. 5. But right here, it can be said that the first objective of the research has been achieved successfully.

Impact Analysis 5

5.1 Prelude

The second objective of the present study is *to assess the relative contributions of climate variability and human activities to cause the alteration of runoff in the South Koel River Basin.* The preceding chapter demonstrated that the rainfall series exhibits no significant trend or step change, whereas the runoff series reveals both gradual (trend) and abrupt (step) changes. This indicates that climate variability has played no role in driving runoff alteration, and therefore its contribution is expected to be insignificant. Consequently, it becomes imperative to examine how human activities particularly land use and land cover transformations have shaped the basin's hydrological behavior. In this chapter, a hydrological modeling approach is applied to disentangle and quantify the respective impacts of climate variability and land use/land cover changes.

5.2 Database and Methods

5.2.1 An Overview of the Methodology

The methodology was designed for the purpose of separating out the relative contribution of climate variability and that of human activities to runoff alteration in the river basin under investigation. The annual runoff time series for the whole study period (1981–2018) was partitioned into a reference period and an interference period based on the change point (year), which was detected by step change analysis. The earlier part, i.e., the period before the change point, is considered as the reference period, while the latter part of the study period, i.e., the period after the change point, is assumed as the interference period

S. Mondal and R. K. Paul, *Climate Variability and Human Interference*, Deltas of the World, https://doi.org/10.1007/978-3-032-13257-4_5

because the magnitude of human interference into any river basin usually increases with the passage of time.

Then SWAT model was calibrated and run with LULC and the weather dataset of the reference period. This model run was named as 'Sim1', which stands for 'Simulation 1'. Another SWAT model was then calibrated and run with the LULC and weather dataset of interference period, and this model run is designated as 'Sim2'. The LULC and weather datasets both were different between the Sim1 and Sim2 models. So, any difference in mean simulated runoff between Sim2 and Sim1 would have been the difference due to both the climate variability and human interference.

Again, the SWAT model, which was calibrated with reference period LULC, was run with the interference period weather dataset, and this model run was designated as 'Sim3'. The LULC in Sim1 and Sim3 were the same, but the weather dataset was different. So, any difference in mean simulated runoff between Sim3 and Sim1 would have been due to variation in climate variability. Again, in between Sim2 and Sim3, the LULC was different, but the weather dataset was the same. So, any difference in mean simulated runoff between Sim2 and Sim3 would have been due to human interference.

After the model run, whatever the calculation was done arithmetically in order to separate out the relative contribution of climate variability and human interference, it was done totally based on the sample dataset. But here, the basic intention is to find the features of the population. So, to infer the population features based on sample data, hypothesis testing was done by applying an independent-samples t-test. The test was applied to know whether there is any significant difference in the population of simulated runoff between (i) Sim2 and Sim1, (ii) Sim3 and Sim1, and (iii) Sim2 and Sim3. After the t-test, Cohen's *d* was computed to know the effect size, and it was computed only when the t-Test result became significant (Fig. 5.1).

Before applying the t-test, the required assumption of normality was checked properly for the concerned variable under investigation, and fortunately, the assumption was not violated for any variables for which the concerned test was applied.

5.2.2 Determining the Reference Period and Interference Period

The whole study period 1981–2018 was divided into the reference period and interference period based on change point detected in the annual runoff series by step change analysis which was carried out in previous chapter. Buishand U test and Buishand Range test detected a common change point in the year 2001, whereas Standard Normal Homogeneity Test (SNHT) detected 2008 as the change point (see Table 4.8). Further analysis was carried out based on both the change points. Hence, the reference and interference period was defined twice, first based on 2001 and then based on 2008. The year 2001 divided the whole study period into the reference period from 1981–2001 and the interference period from 2002–2018, whereas based on 2008 as the change point, the period 1981–2008 was

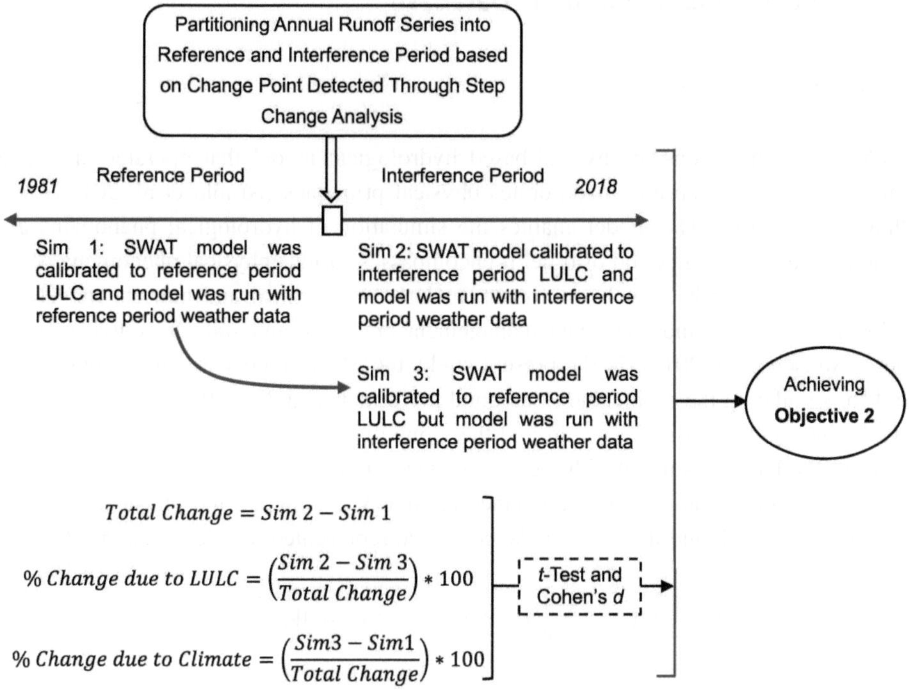

Fig. 5.1 Methodology for assessing the relative contribution of climate variability and human activities to runoff alteration

identified as the reference period and the period 2009–2018 was defined as interference period (Table 5.1).

Table 5.1 Identification of the reference period and interference period

Sl. No.	Change point	Period	Time-step
1	2001	Reference period	1981–2001
		Interference period	2002–2018
2	2008	Reference period	1981–2008
		Interference period	2009–2018

Source Computed by the authors

5.2.3 SWAT Model and Input Database

(a) *About SWAT Model*

SWAT is a comprehensive physical-based hydrological model that operates on a continuous daily time step and incorporates physical principles (Arnold et al. 2012; Linlin Zhang et al. 2020). The model enables the simulation of hydrological phenomena at a basin level over a long period of time. It encompasses various physical phenomena occurring within a watershed, such as climate, hydrology, chemical and nutrient movement, sediment transport, land cover and management practices, and water resource management (Neitsch et al. 2011). In the present study, the SWAT model was incorporated as a tool for simulating the runoff due to its good performance ability, efficient computational capabilities, and extensive usage around the globe in runoff simulation applications.

The SWAT model simulates the runoff (surface runoff) based on the SCS-CN method. However, it utilizes a daily water balance approach to simulate the physical processes within a watershed, and this water balance can be represented as (Neitsch et al. 2011):

$$SW_t = SW_O + \sum_{i=1}^{t}\left(R_{day} - Q_{surf} - E_a - W_{seep} - Q_{gw}\right) \tag{5.1}$$

where the final soil water content (SW_t) is equal to the initial soil water content (SW_O) on the day i, plus the cumulative sum of various components over the simulated days (t), including precipitation (R_{day}), surface runoff entering the vadose zone (Q_{surf}), evapotranspiration (E_a), seepage from the soil profile (W_{seep}), and water return flow (Q_{gw}) (all measured in mm H_2O). This equation allows SWAT to calculate the changes in soil water content and trace the movement of water throughout the watershed.

(b) *Preparing SWAT Input Database*

In the study, the ArcSWAT 2012.10.5.24 interface within ArcMap 10.5 was utilized to set up and run the SWAT model for simulating the annual runoff. Data such as elevation, soil, land use and land cover, weather datasets, and runoff data were collected from the various authentic sources and fed into the SWAT model (Table 5.2). The soil datasets were resampled to a spatial resolution of 30 m to maintain computational efficiency.

(i) *Elevation Database*

The SRTM DEM was employed to calculate the flow direction and flow accumulation, extract the stream networks, and delineate the watershed. Additionally, a slope map

Table 5.2 SWAT input database

Sl. No.	Input data	Source	Spatial resolution
1	Elevation data	SRTM data from USGS Earth Explorer https://earthexplorer.usgs.gov (USGS Earth Explorer 2022)	30 m
2	Soil data	Soil grid data from https://soilgrids.org (ISRIC-WDC 2020)	250 m (resampled to 30 m)
3	LULC data	Prepared by classifying TM and OLI images from USGS Earth Explorer, https://earthexplorer.usgs.gov (USGS Earth Explorer 2022)	30 m
4	Weather data (precipitation, maximum and minimum temperature, relative humidity, wind speed, and solar radiation)	NASA Power Data Access, Langley Research Center, https://power.larc.nasa.gov/data-access-viewer/) (NASA 2020)	0.625° × 0.50°

Source Computed by the authors

was generated from the DEM image using ArcSWAT interface for subsequent runoff simulation analysis.

(ii) *Soil Database*

The soil datasets were downloaded from the https://soilgrids.org website (ISRIC-WDC 2020) and formatted as per the SWAT model's specifications. The SWAT model required a set of input variables, some of which were readily available, while others needed to be derived using pedo-transfer functions (PTFs) (Table 5.3). However, the available water capacity of the soil layer (SOL_AWC) is calculated using Eqs. 5.2 and 5.3 (Dijkerman 1988):

$$\theta_{33} = 0.3697 - 0.0035 \times S \tag{5.2}$$

$$\theta_{1500} = 0.0074 + 0.0039 \times C \tag{5.3}$$

where SOL_AWC is the difference between $\theta_{33} - \theta_{1500}$ measured in cm cm^{-1}, θ_{33} is the soil water content at field capacity, θ_{1500} is the soil water content at wilting point, S is the % sand, and C is the % clay.

Similarly, the soil hydraulic conductivity (SOL_K) in cm day^{-1} (Cosby et al. 1984) and moist soil albedo (SOL_ALB) (Wang et al. 2005) are calculated using Eqs. 5.4 and 5.5 respectively.

Table 5.3 List of required soils variable for SWAT

Sl. No.	Variable's name	Abbreviation	Remarks
1	Soil hydrologic group	HYDGRP	1. Required only for the SWAT ArcView Interface
			2. Available (250 m soil grid)
2	Maximum rooting depth of soil profile (mm)	SOIL_ZMX	Available (250 m soil grid)
3	Depth from soil surface to bottom of layer (mm)	SOL_Z	Available (250 m soil grid)
4	Moist bulk density (mg/m^3 or g/cm^3)	SOL_BD	Available (250 m soil grid)
5	Available water capacity of the soil layer (mm H_2O/mm soil)	SOL_AWC	Need to prepare/calculate
6	Saturated hydraulic conductivity (mm/hr)	SOL_K	Need to prepare/calculate
7	Organic carbon content (% soil weight)	SOL_CBN	Available (250 m soil grid)
8	Clay content (% soil weight)	SOL_CLAY	Available (250 m soil grid)
9	Silt content (% soil weight)	SOL_SILT	Available (250 m soil grid)
10	Sand content (% soil weight)	SOL_SAND	Available (250 m soil grid)
11	Rock fragment content (% total weight)	SOL_Rock	Available (250 m soil grid)
12	Moist soil albedo (top layer)	SOL_ALB	Need to prepare/calculate
13	USLE equation soil erodibility (K) factor (units 0.013 (metric ton m^2 hr)/(m^3-metric ton cm)) (top layer)	USLE_K	Need to prepare/calculate

Source Tabulated by the authors

$$SOL_K = 60.96 \times 10^{(-0.6+0.0126 \times S-0.0064 \times C)} \tag{5.4}$$

$$SOL_{ALB} = 0.1807 + 0.1019 \times \exp(-3.53 \times \theta_{33}) \tag{5.5}$$

where C is the % clay, S is the % sand, and θ_{33} is the soil water content at field capacity.

To calculate the USLE (Universal Soil Loss Equation) soil erodibility factor (K) in cm day^{-1}, Williams (1995) proposed the following PTFs:

$$KUSLE = E_S \times E_{C-T} \times E_{OC} \times E_{HS} \tag{5.6}$$

$$E_S = 0.2 + 0.3 \times \exp[-0.256 \times S \times (1 - T \div 100)] \tag{5.7}$$

$$E_{C-T} = [T/(C+T)^{0.3}] \tag{5.8}$$

$$E_{OC} = [1 - (0.25 \times OC/(OC + \exp(0.72 - 2.95 \times OC))] \quad (5.9)$$

$$E_{HS} = 1 - \{0.7 \times (1 - S \div 100) \div [(1 - S \div 100) + \exp(-5.51 + 22.9 \times (1 - S \div 100))]\} \quad (5.10)$$

where S is the % sand, T is the % silt, C is the % clay, and OC is the % organic carbon.

At last, a soil textural map was prepared based on the USDA classification method (Soil Survey Division Staff 1993), and three textural classes have been identified, i.e., sandy clay loam, clay loam, and clay. For that, first, it needs to normalize the data layers of sand, silt, and clay individually using the following equation:

$$Weighted\ average = \frac{(Sand\_0\_cm + w_1) + (Sand\_5\_cm + w_2) + \cdots n}{w_1 + w_2 + w_3 + w_4 + w_5 + \cdots n} \quad (5.11)$$

Then, the soil textural class has been delineated using the following USDA equation

$$Sandy\ clay\ loam = Clay >= 20\%\ and\ Clay < 35\%\ and\ Silt < 28\%\ and\ Sand > 45\% \quad (5.12)$$

$$Clay\ loam = Clay >= 27\%\ and\ Clay < 40\%\ and\ Sand >= 20\%\ and\ Sand <= 45\% \quad (5.13)$$

$$Clay = Clay >= 40\%\ and\ Sand <= 45\%\ and\ Silt < 40\% \quad (5.14)$$

(iii) *LULC Database*

The LULC maps were prepared using QGIS 3.28.4 based on the random forest classification scheme for the years 1995 and 2016. It can be mentioned that the LULC map for the year 1995 was taken as a representative of the reference period's LULC, while the LULC map for the year 2016 was a representative of the interference period's LULC. However, five LULC classes were identified, namely forest, agricultural land, urban built-up area, water bodies, and bare ground from the Landsat satellite images (Table 5.4).

The accuracy assessment of the prepared LULC maps was done using the confusion matrix (user's, producer's, and overall accuracies) and Kappa statistics methods in ERDAS Imagine 2015. A confusion matrix organizes the known LULC types of reference data into columns and the classified data into rows (Banko 1998). Producer's accuracy determines the ratio of correctly classified pixels for a specific LULC class to the total number of reference pixels for that class. User's accuracy measures the ratio of correctly classified pixels for a particular class to the total number of pixels classified as that class in the classified map. Overall accuracy provides an overall measure of accuracy for the entire

Table 5.4 Description of the satellite image

Name of the satellite	Sensor	Path	Row	Time of acquisition	Time of downloading
Landsat-05	TM	140	44	Nov, 1995	6/10/2022
		140	45		
		141	44		
Landsat-08	OLI	140	44	Nov, 2016	6/10/2022
		140	45		
		141	44		

Source Computed by the authors

classification by calculating the ratio of correctly classified pixels of all classes to the total number of pixels in the study area.

Other than the confusion matrix, the Kappa index of agreement, also known as Cohen's kappa coefficient, is a statistical measure that was widely used to assess the agreement between the classified map and reference data (Nguyen et al. 2020). It considers the agreement expected by chance alone, offering a robust accuracy measure. The Kappa index ranges from 0 to 1, where 1 represents perfect agreement, and 0 indicates no agreement. Commonly used thresholds for interpreting Kappa values are as follows: Kappa < 0.20 indicates poor agreement, 0.20 ≤ Kappa < 0.40 indicates fair agreement, 0.40 ≤ Kappa < 0.60 indicates moderate agreement, 0.60 ≤ Kappa 0.80 indicates substantial agreement, and Kappa ≥ 0.80 indicates almost perfect agreement.

Following equations are used for carrying out the accuracy assessment.

$$PA = \frac{(\sum NT_c)}{(\sum NT_{rp})} \times 100 \tag{5.15}$$

$$UA = \frac{(\sum NT_c)}{(\sum NT_{cp})} \times 100 \tag{5.16}$$

$$OA = \frac{\sum_{i=1}^{nc} e_{ii}}{NT} \times 100 \tag{5.17}$$

$$\overset{\vee}{K} = \frac{\left(NT \times \sum_{i=1}^{nc} e_{ii}\right) - Pe}{(NT)^2 - Pe} \times 100 \tag{5.18}$$

$$NT = \sum_{i=1}^{nc} \sum_{j=1}^{nc} e_{ij} \tag{5.19}$$

where PA = Producer's Accuracy, NT_c = number of correctly classified pixels for a specific class, NT_{rp} = number of reference pixels for that class, UA = User's Accuracy, NT_{cp} = number of classified pixels for that class, OA = overall accuracy, nc = total number of classes, e_{ii} = element in ith row and ith column, NT = total number of samples, $\overset{\vee}{K}$ = Kappa index, e_{ij} = element in ith row and jth column, and Pe = chance agreement.

(iv) *Weather Database*

Daily precipitation, maximum and minimum temperature, relative humidity, wind speed, and solar radiation data for the years 1981–2018 were downloaded from the NASA Power Data Access, Langley Research Centre (NASA 2020) and prepared as per the SWAT format using weather generator (WGN) parameters estimation tool which was downloaded from https://swat.tamu.edu/software/ website. The missing weather data were defined by using the SWAT model's default value of − 99.

(v) *Runoff Database*

Daily runoff data from 1973–2018 was acquired from the Central Water Commission, The Department of Water Resources, Ministry of Jal Shakti, Government of India (https://indiawris.gov.in/wris/#/) for the Jarikela gauge station, which was used as the outlet in the watershed delineation process. The runoff datasets were used for model validation and calibration purposes.

However, the entire South Koel River basin was divided into 27 subbasins by using a flow accumulation threshold value of 2% basin area (206 km^2). The user selects the outlet to generate the subbasins, which are defined using DEM in a process like watershed delineation. Sub-basins are spatially distributed, and the streamflow occurs from one sub-basin to another through routing properties established within the model (Kalcic et al. 2015). The model automatically further subdivided the subbasins into Hydrologic Response Units (HRUs) by grouping together areas with similar land use, soil type, and slope characteristics based on the predefined thresholds for each category. The number of HRUs can be adjusted by the user, but for this study, the SWAT-generated HRUs remained unchanged to maintain computational efficiency.

5.2.4 Calibration and Validation

Calibration refers to the adjustment of the model's parameters by comparing model output with real-world observations, ensuring consistent responses over time. In the study, the model parameters were modified until the simulated runoff was closely matched with the observed runoff. On the other hand, validation is the method of authenticating the model output. Here, the model is run with already calibrated parameters and by using an independent dataset that was not used for calibration, and then model output is compared with real-world output. If the model fails to produce authentic output in the validation stage, it is recalibrated and validated again until the satisfactory model performance is achieved.

In this study, SWAT-CUP software was applied to assess the prediction uncertainty of the SWAT model through calibration and validation. Its interface is designed to combine the multiple calibration and uncertainty analysis methods for the SWAT model, which include SUFI-2, PSO, GLUE, ParaSol, and MCMC algorithms.

Calibration can be done manually, semi-automatically, and automatically. In this study, the SUFI-2 algorithm was utilized in the SWAT-CUP application program, which can perform both automatic and semi-automatic calibration of the model. It served the purpose of analysing sensitivity and uncertainty, as well as calibrating and validating numerous parameters in the SWAT model (Thavhana et al. 2018).

Considering the potential for significant uncertainty and inaccuracies in hydrologic models, the SUFI-2 algorithm assumes a Gaussian distribution for input parameter uncertainty, with output uncertainty calculated at the 95PPU (95% prediction uncertainty) level (Szcześniak and Piniewski 2015). The study involved running five iterations, each consisting of 500 simulations, with different parameter values sampled in the parameter space until optimization was achieved. During execution, SWAT-CUP employs Latin hypercube sampling of the parameter space, and the post-processing involves computing the NSE for the watershed (Szcześniak and Piniewski 2015). After each iteration in the SWAT-CUP optimization process, the parameter ranges are narrowed down based on the suggested ranges, and the next iteration begins with the objective of improving the fitness of the SWAT model. Parameters can be adjusted in three ways: replacement, relative adjustment (as a percentage), and absolute adjustment (adding or subtracting from the current value).

The model's calibration and validation were judged using the two objective functions, namely Nash–Sutcliffe Efficiency (NSE) and Coefficient of Determination (R^2), which are the most popular and widely used techniques. The NSE referred to as the efficiency index (E_f) (McCuen et al. 2006), is calculated using the formula:

$$E_f = 1 - \frac{\sum^n (\widehat{Y_i} - Y_i)^2}{\sum (Y_i - \overline{Y})^2} \tag{5.20}$$

where $\widehat{Y_i}$ represents predicted values, $\widehat{Y_i}$ represents measured values, $\overline{Y}$ represents the average of the measured values, and n represents the sample size of runoff values. The Nash–Sutcliffe numbers can range from $-\infty$ to 1.0, with 1.0 indicating a perfect match between modelled runoff and observed runoff. A value greater than 0.5 is generally accepted as indicative of a sufficiently calibrated model (Nash and Sutcliffe 1970). During calibration, model parameters are adjusted to maximize NSE, aiming for the highest possible value of 1.

On the other hand, the R^2 value ranges from 0 to 1, where 0 indicates mismatch and 1 represents a perfect match between observed runoff and model-generated runoff. It can

Table 5.5 Runoff calibration and validation periods

Sl. No.	Change point	Period	Warm up period	Calibration period (monthly)	Validation period (monthly)
1	2001	Reference period of 1981–2001	1981–1983	1984–1999	2000–2001
2		Interference period of 2002–2018	2002–2004	2005–2016	2017–2018

Source Computed by the authors

be mentioned that till now, no absolute criteria for judging the model performance have been firmly established in the literature.

However, the calibration and validation period which were used in the present study is defined in Table 5.5.

5.2.5 Sensitivity Analysis

Global sensitivity analysis is automatically performed during the calibration process in SWAT-CUP to determine the sensitivity of parameters. This analysis involves regressing the Latin hypercube-generated parameters against objective function values, resulting in multiple regression systems (Sao et al. 2020). In SUFI-2, the sensitivity of parameters was assessed using t-statistic values, with larger absolute t-statistic values indicating greater sensitivity. The significance of sensitivity is determined by p-values, where a parameter is considered significant if its p-value approaches zero. These sensitivities represent the average changes in the objective function caused by variations in each parameter while keeping all other parameters constant. These sensitivities are relative and based on linear approximations. In the study, a total of 16 SWAT input parameters were employed for sensitivity analysis, which was selected from the study of Arnold et al. (2012) and Visakh and Raju (2018).

5.2.6 SWAT Run and Analysis of Output

(a) *Separating* the *Impact of Climate Variability and Human Activity*

As mentioned earlier, three separate model simulations (Sim) are required to separate out the contribution of climate variability and that of human activity (LULC). Sim1 involves the model calibrated to reference period LULC and the model run with reference period weather datasets. Sim2, on the other hand, involves the model calibrated to interference period LULC and the model run with interference period weather datasets. Sim3 involves the model calibrated to reference period LULC, but the model runs with interference period weather datasets.

The difference between Sim2 and Sim1 is caused by the impact of both climate variability and LULC change. The difference between Sim3 and Sim1, on the other hand, is due to the impact of climate variability only since, here, LULC remains the same in both simulations, but the weather dataset is different. The difference between Sim2 and Sim3 is caused by the change in LULC since, here, the weather dataset remains the same but LULC changes.

The arithmetic for computing the contribution of LULC change and climate variability, as discussed above, are mentioned below in the form of equations:

$$Total\ change = Sim2 - Sim1 \tag{5.21}$$

$$\%\ Change\ due\ to\ Climate\ Variability = \left(\frac{Sim3 - Sim1}{Total\ Change}\right) * 100 \tag{5.22}$$

$$\%\ Change\ due\ to\ LULC = \left(\frac{Sim2 - Sim3}{Total\ Change}\right) * 100 \tag{5.23}$$

(b) *Independent Sample t-Test*

An Independent t-Test was applied to know whether there was a significant difference in mean simulated runoff between Sim2 and Sim1, between Sim3 and Sim1, and between Sim2 and Sim3. The significance difference provides evidence that whatever contribution is computed for climate variability and human activity based on sample data could be equally applicable in their population. The effect size was computed by Cohen's *d*, and it was computed only when the t-Test results were found significant. It is worth mentioning that the t-Test assumes the sample data to be normally distributed. So, before applying the t-Test, the simulated runoff of each simulation phase was checked for normality (see Sect. 4.2.2).

The test statistic for the independent sample t-test, as propounded by Sir William Gosset under the pseudonym "Student" in 1908 (Drummond and Tom 2011), is computed by dividing the mean difference by their standard error, as mentioned below.

$$t = \frac{\overline{x}_1 - \overline{x}_2}{SE_{\overline{x}_1 - \overline{x}_2}} \tag{5.24}$$

The standard error of mean difference denoted as $S_{\overline{x}_1 - \overline{x}_2}$, is computed as follows:

$$SE_{\overline{x}_1-\overline{x}_2} = S_p\sqrt{\frac{1}{n_1}+\frac{1}{n_2}} \tag{5.25}$$

S_p is the estimated pooled standard deviation which is computed as follows:

$$S_p = \sqrt{\frac{(n_1-1)s_1^2+(n_2-1)s_2^2}{n_1+n_2-2}} \tag{5.26}$$

where $\overline{x}_1$ is the mean value of the first sample group, $\overline{x}_2$ is the mean value of the second sample group, n_1 is the size of observations in the first sample group, n_2 is the size of observations in the second sample group, s_1^2 is the variance of the first sample group, and s_2^2 is the variance of the second sample group.

The *p*-value of the test statistic is computed by applying t-Distribution since the concerned statistic follows this distribution, and the shape of the distribution curve depends upon the degree of freedom (*df*), which is defined as:

$$df = n_1 + n_2 - 2 \tag{5.27}$$

(c) *Cohen's d*

Cohen's *d* was computed for those samples which have significant mean differences in runoff across the simulations. It is a kind of effect size of the difference (Cohen 1992), and it is computed by taking the difference between two means and dividing it by a standard deviation. The formula for Cohen's *d* is as follows:

$$d = \frac{\overline{x}_1-\overline{x}_2}{s} \tag{5.28}$$

In this equation, s represents the pooled standard deviation, which is determined as follows (Cohen 1988)

$$s = \sqrt{\frac{(n_1-1)s_1^2+(n_2-1)s_2^2}{n_1+n_2-2}} \tag{5.29}$$

where s_1^2 is the variance of the first sample group, and s_2^2 is the variance of the second sample group. The variance for the first group is calculated as

$$s_1^2 = \frac{1}{n_1-1}\sum_{i=1}^{n_1}(x_{1,i}-\overline{x}_1)^2 \tag{5.30}$$

Similarly, the variance for the second group can be calculated. It should be noted that Cohen's *d* provides a standardized measure of the effect size, representing the magnitude

Table 5.6 Magnitude of Cohen's *d*

Sl. No.	Values of Cohen's *d*	Remarks
1	0.01	Very small
2	0.2	Small
3	0.5	Medium
4	0.8	Large
5	1.2	Very large
6	2	Huge

Source Tabulated by the authors

of the difference between the means in relation to the variability within the data. The below table shows the descriptions for different magnitudes of Cohen's *d* ranging from 0.01 to 2.0. These descriptors were initially proposed by Cohen (1988) and later expanded upon by Sawilowsky (2009) (Table 5.6).

5.3 Results and Discussion

5.3.1 Results of the Model's Calibration and Validation

In each reference or interference period, the first three years were used for warm up, the last two years were used for validation, and the time period in between was used for calibration. The same break up for warm up period, calibration, and validation period was followed for both 2001-based and 2008-based defined reference and interference periods. The fitted parameters of Sim1 were applied to the Sim3 model. However, a total of 16 parameters were chosen in order to calibrate the models. Their respective ranges and fitted values are presented in below (Table 5.7):

The goodness of model fit was judged by the NSE and R^2. These tests were performed in both phases, at the time of calibration and at the time of validation too. During calibration, the monthly average simulated runoff closely matched the observed values for all periods. Based on the 2001 change point, the simulated runoff was 196.11 cumec for the reference period 1984–1999, while the observed runoff was 159.41 cumec, and for the interference period 2005–2016, the simulated runoff was 108.11 cumec and the observed runoff was 95.19 cumec. Based on the 2008 change point, the simulated runoff was 198.30 cumec for the reference period 1984–2006, with the observed runoff at 150.60 cumec, and for the interference period 2012–2016, the simulated runoff was 84.16 cumec and the observed runoff was 72.80 cumec.

The NSE values ranged from 0.75–0.83. The values indicate greater model efficiency. The highest NSE value was observed for the period of 1984–1999 (0.83) and the lowest for 2005–2016 (0.75). Similarly, the R^2 value ranges from 0.77–0.86. The best fitting is

Table 5.7 Results of SWAT calibration parameters with absolute ranges and fitted values

Sl. No.	Parameters with extension	Range	Fitted value				Description
			Reference period (1981–2001)	Interference period (1981–2008)	Reference period (2002–2018)	Interference Period (2009–2018)	
1	CN2.mgt	– 0.4	0.07	0.07	– 0.15	– 0.15	SCS runoff curve number
2	ALPHA_BF.gw	0–1	0.84	0.84	0.98	0.98	Baseflow alpha factor (days)
3	GW_DELAY.gw	30–450	342.9	342.9	304.05	304.05	Groundwater delay (days)
4	GWQMN.gw	0–2	0.77	0.77	1.71	1.71	Threshold depth of water in the shallow aquifer required for return flow to occur (mm)
5	GW_REVAP.gw	0–0.2	0.02	0.02	0.12	0.12	Groundwater "revap" coefficient
6	ESCO.hru	0.8–1	0.9	0.9	0.83	0.83	Soil evaporation compensation factor
7	CH_N2.rte	0–0.3	0.1	0.1	0.21	0.21	Manning's "n" value for the main channel
8	CH_K2.rte	5–130	86.88	86.88	7.81	7.81	Effective hydraulic conductivity in main channel alluvium
9	ALPHA_BNK.rte	0–1	0.61	0.61	0.13	0.13	Baseflow alpha factor for bank storage
10	SOL_AWC.sol	– 0.6	0.28	0.28	0.05	0.05	Available water capacity of the soil layer
11	SOL_K.sol	– 1.6	0.01	0.01	0.09	0.09	Saturated hydraulic conductivity
12	SOL_BD.sol	– 1.1	– 0.3	– 0.3	– 0.43	– 0.43	Moist bulk density
13	SURLAG.bsn	0.05–24	7.35	7.35	23.94	23.94	Surface runoff lag time
14	OV_N.hru	0.01–30	5.86	5.86	26.33	26.33	Manning's "n" value for overland flow

(continued)

Table 5.7 (continued)

Sl. No.	Parameters with extension	Range	Fitted value				Description
			Reference period (1981–2001)	Interference period (1981–2008)	Reference period (2002–2018)	Interference Period (2009–2018)	
15	REVAPMN.gw	0–500	437.5	437.5	421.25	421.25	Threshold depth of water in the shallow aquifer for "revap" to occur (mm)
16	RCHARG_DP.gw	0–1	0.21	0.21	0.7	0.7	Deep aquifer percolation fraction

Source Computed by authors

found for the period of 1984–1999 (0.86). However, the results of each period indicate that the models exhibit excellent efficiency and a remarkable fit for real-world observations (Table 5.8).

The 95 PPU plot evaluated the performance of the calibrated model by comparing the observed data against the model's predictions based on the 95th percentile prediction uncertainty (PPU). A good calibration would result in most observed data points falling within the 95 PPU range, indicating a reasonable level of prediction certainty. The results indicate that the observed runoff data fell within the 95 PPU range, which indicated very

Table 5.8 Results of runoff calibration and validation periods

Change point	Warm up period	Calibration period	Validation period	Monthly average runoff (cumec)		R^2	NSE
				Simulated	Observed		
2001	1981–1983	Reference period calibration: 1984–1999	–	196.11	159.41	0.86	0.83
	2002–2004	Interference period calibration 2005–2016	–	108.11	95.19	0.77	0.75
2008	1981–1983	Reference period calibration 1984–2006	–	198.30	150.60	0.82	0.77
	2009–2011	Interference Period Calibration 2012–2016	–	84.16	72.80	0.79	0.77
2001	–	–	Reference period validation 2000–2001	172.97	144.43	0.90	0.88
	–	–	Interference period validation 2017–2018	86.55	85.27	0.93	0.91
2008	–	–	Reference period validation 2007–2008	220.15	168.55	0.81	0.76
	–	–	Interference period validation 2017–2018	84.16	72.80	0.79	0.77

Source Computed by the authors

good prediction accuracy (Fig. 5.2). The comparison of simulated runoff and observed runoff is shown as follows (Figs. 5.3, 5.4, 5.5 and 5.6).

5.3.2 Results of Sensitivity Analysis

During the calibration process, the global sensitivity analysis was performed in the SWAT-CUP environment, which identified the relative sensitivity parameters of the model's calibration (Table 5.9). It can be mentioned that the sensitivity analysis parameters were adjusted in a sequence based on their relative sensitivity within their specified ranges until the simulated runoff was adjusted to the required quantity. This iterative process was repeated until the simulated runoff was obtained within a range of $\pm 15\%$ compared to the observed runoff. The results show that the most sensitive parameters for the simulation period of 1984–1999 are ALPHA_BNK (0.000), CN2 (0.001), and CH_N2 (0.003), respectively. On the other hand, for the simulation period of 2005–2016, CN2 (0.05) is the most sensitive parameter among the other parameters.

5.3.3 An Overview of Climate Variability and LULC Change

Rainfall is the most significant climatic parameter from a hydrologic perspective since it is the only causal factor that produces runoff in the basin under investigation. Hence, rainfall was taken here as the representative of the climate. So, climate variability was evaluated by evaluating the rainfall only. That is why step change detection and trend analysis was performed only for the rainfall instead of applying the same analysis for all other climate parameters like temperature, humidity, etc.

The results of rainfall variability showed that there was no significant step change or trend in the rainfall data series. After partitioning the whole rainfall series (1981–2018) into subseries, the trend analysis was done once again for the sub-series, and it was observed the pre-monsoon series showed an increasing trend from 1991 onward, whereas the winter series showed decreasing trend up to 1993 (Table 4.15). However, these type of changes during pre-monsoon and winter does not affect the runoff generation mechanism, and this mechanism gets affected only due to changes in the monsoon rainfall series. But it was not possible to partition monsoon rainfall series based on the eye-estimation break point analysis. So, whatever variability is observed in rainfall is not important from a hydrologic perspective.

However, these changes in pre-monsoon and winter rainfall do not significantly influence runoff generation, which is primarily affected by monsoon rainfall. It was not possible to identify break points in the monsoon rainfall series using eye-estimation, so the observed variability in rainfall is considered hydrologically insignificant.

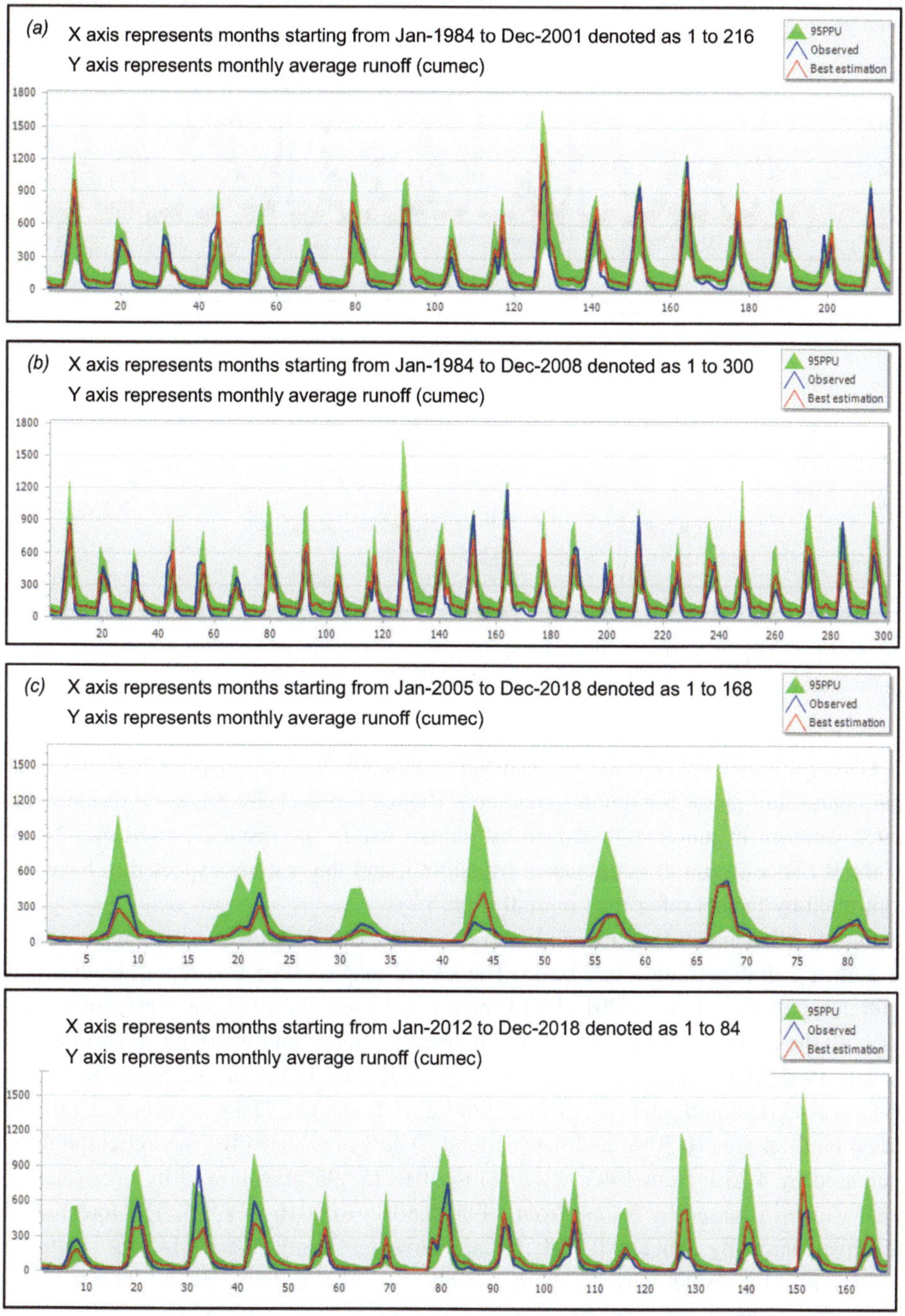

Fig. 5.2 Calibration results of the SWAT model obtained using SWAT-CUP, showing observed and simulated streamflow along with the 95% prediction uncertainty (95PPU) band from **a** Jan-1984 to Dec-2001, **b** Jan-1984 to Dec-2008, **c** Jan-2005 to Dec-2018, and **d** Jan-2012 to Dec-2018. *Source* Computed based on CWC discharge data, 2020

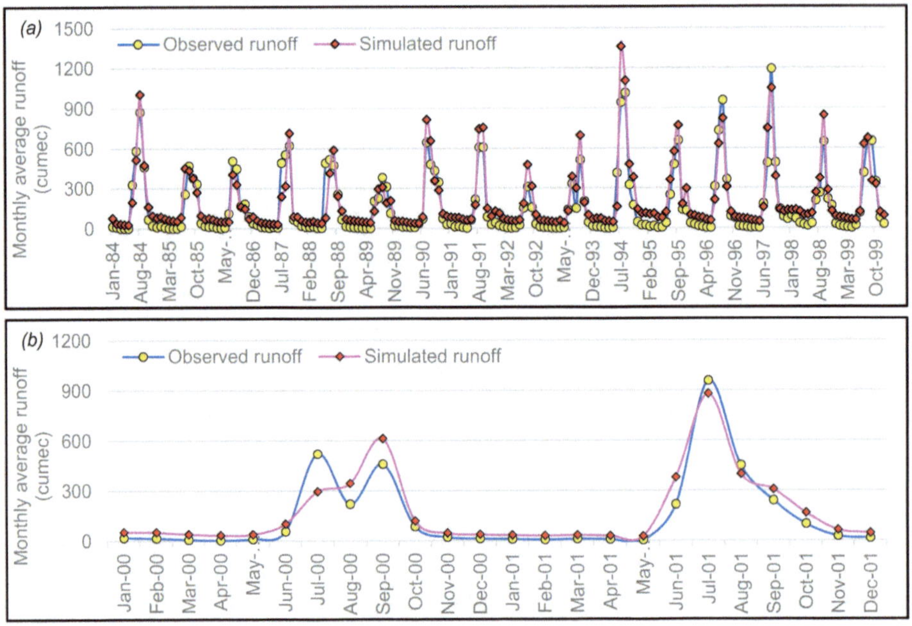

Fig. 5.3 **a** Calibration (1984–1999) and **b** validation (2000–2001) of monthly average observed runoff and simulated runoff. *Source* Computed based on CWC discharge data, 2020

Here, an attempt was made to evaluate statistically the role of rainfall as the causal and controlling factor for runoff generation. Regression analysis was done to know how much variance in runoff is explained by rainfall itself. The results showed that 21.75% of the variance in runoff is explained by rainfall, and the rest are expected to have been controlled by factors other than rainfall (Fig. 5.7).

On the other hand, land use and land cover (LULC) was taken as the representative of human activities in the study basin. The LULC map of 1995 was representative of the reference periods, while the 2016 LULC map was representative of the interference period for both the 2001-based and 2008-based defined reference and interference periods.

The 1995 LULC map shows that the most dominant LULC in the South Koel River basin was agricultural land (53.70%), followed by forest (44.27%), waterbodies (1.17%), urban built-up area (0.50%) and bare ground (0.36%) respectively. The agricultural land increased by 4.25% from 1995 (53.70%) to 2016 (57.95%), followed by urban built-up area, which increased by 3.62% from 1995 (0.50%) to 2016 (4.12%). The forest shows negative change by about − 0.17%. Its area coverage changed from 44.27% in 1995 to 37.23% in 2016. Waterbody and bare ground also show negative growth between 1995 and 2016 − 0.66% and − 0.17%, respectively (Fig. 5.8 and Table 5.10).

The accuracy assessment report is also satisfactory. The producer's accuracy for both maps is above 80% for all classes except for bare ground, for which it was 76% in 1995

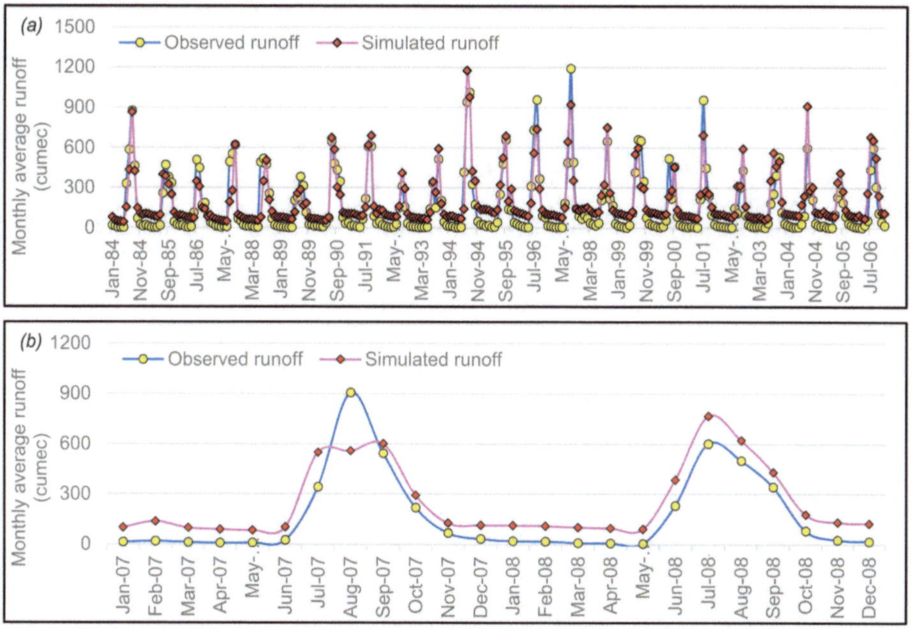

Fig. 5.4 **a** Calibration (1984–2006) and **b** validation (2007–2008) of monthly average observed runoff and simulated runoff. *Source* Computed based on CWC discharge data, 2020

and 66.67% in 2016. The user's accuracy is also above 80% for all the classes. The overall accuracy for both the LULC maps is again more than 88%, and it is 88.73 for the 1995 map and 90.20 for the 2016 map. The conditional Kappa statistics are well above 80% for all the classes except for the built-up area in 1995, for which it is 79.30. The overall Kappa statistics for both maps are more than 85%, and it is 85.59 for 1995 and 87.28 for the 2016 map (Table 5.11).

5.3.4 Relative Contribution of Climate Variability and LULC Change

A total of six simulation models were developed to analyse the impact of climate variability and LULC change on runoff alteration. Three models were formulated for reference and interference periods which were defined based on the year 2001 as a change point, and another three were formulated for reference and interference periods which were defined based on the year 2008 as a change point (Table 5.12).

For the change point in 2001, Simulation 1 (Sim1) entails the model calibrated to the reference period (1981–2001) LULC and the model run for the reference period (1981–2001) weather dataset. Sim2 entails the model calibrated to the interference period (2002–2018) LULC and the model run for the interference period (2002–2018) weather dataset.

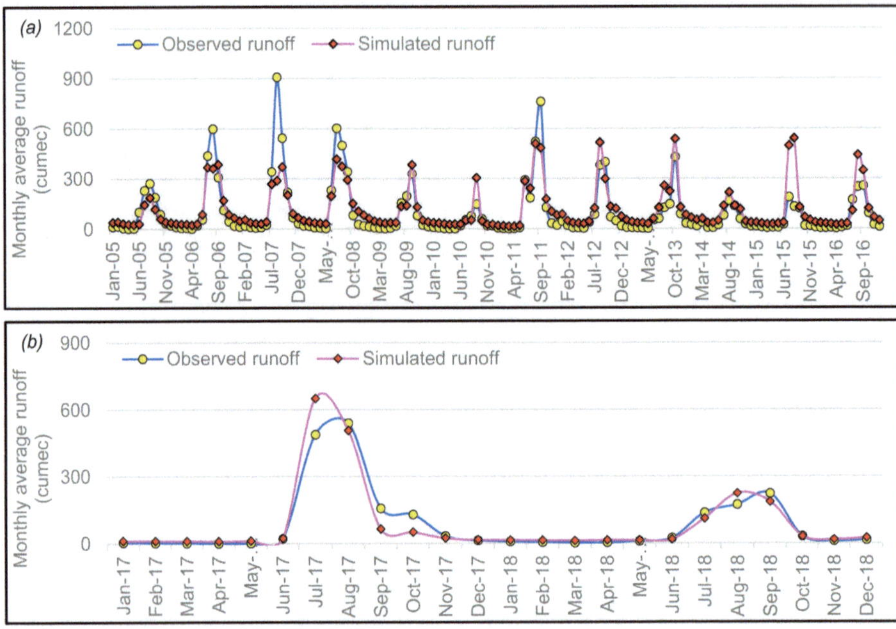

Fig. 5.5 **a** Calibration (2005–2016) and **b** validation (2017–2018) of monthly average observed runoff and simulated runoff. *Source* Computed based on CWC discharge data, 2020

Sim3, on the other hand, entails the model calibrated to the reference period (1981–2001) LULC and interference period weather dataset (2002–2018).

The mean annual runoff generated in Sim1 (1981–2001) was 568.54 mm which was reduced to 372.67 mm in Sim2 (2002–2018), i.e., streamflow decreased in the latter part of the study period in reference to that in the earlier part. Sim3, on the other hand, produced a mean runoff of 652.28 mm, which is much higher than the mean runoff generated in Sim2 and a bit lower than that generated in Sim1, indicating the increase in runoff due to climate variability and the decrease due to land use and land cover (LULC) change.

So, as per the simulation results, the total amount of change in runoff due to both climate variability and LULC = (Sim2–Sim1) = (372.67–568.54) = − 195.87 mm. The % of change in runoff due to climate variability = [(Sim3–Sim1)/Total change] * 100 = [(652.82–568.54)/195.87] = [84.28/195.87] * 100 = 43.03%. The contribution of LULC = [(Sim2–Sim3)/Total change] * 100 = [(372.67–652.82)/195.87] = [− 280.15/195.87] * 100 = − 143.03%.

So, based on the change point in 2001, the annual runoff decreased by 195.87 mm in the reference period (1981–2001) as compared to the interference period (2002–2018). This reduction in runoff was influenced by both climate variability and land use/land cover (LULC) change, which had opposite effects on the alteration of runoff. The impact of climate variability resulted in an increase in runoff by 84.28 mm. On the other hand,

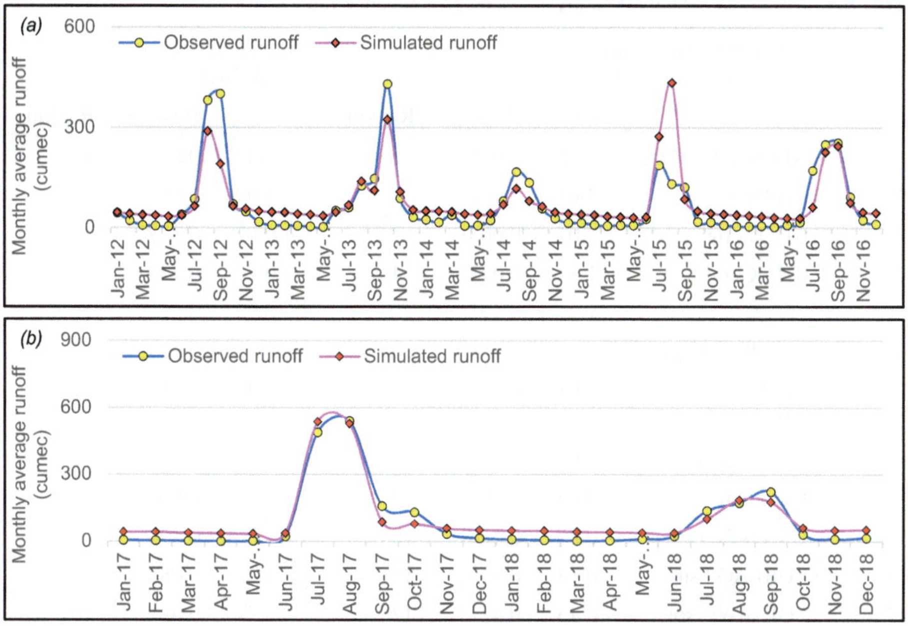

Fig. 5.6 **a** Calibration (2012–2016) and **b** validation (2017–2018) of monthly average observed runoff and simulated runoff. *Source* Computed based on CWC discharge data, 2020

LULC change contributed to a decrease in runoff by 280.15 mm. The combined effect of both the factors therefore resulted in an effective decrease in runoff. In terms of %, the contribution of climate variability accounted for 43.03% in increasing the runoff, while the contribution of LULC change accounted for 143.03% in decreasing the runoff. So, it is evidenced that the change in LULC seems to have a more pronounced effect in reducing runoff compared to the influence of climate variability, which led to a bit increase in runoff during the specified periods (Table 5.13).

Similar features are found while taking 2008 as the change point i.e., runoff increased due to climate change and decreased due to LULC change, and there was a decrease in mean runoff in the interference period as compared to the reference period, which occurred due to combined effect of both climate change and LULC change, although the values are not numerically same (Tables 5.12 and 5.13).

In connection to the year 2008 as the change point, the annual mean runoff was reduced to 224.64 mm in the reference period (1981–2008) as compared to that in the interference period (2009–2018), and this reduction is due to combined effect of climate variability and human intervention but their role is just opposite. The climate variability caused an increase in the runoff by an amount of 58.13 mm, while the LULC change led to the runoff decrease by about 282.77 mm. The rate of contribution of climate variability in

Table 5.9 Results of sensitivity analysis considering change point 2001

Sl. No.	Parameters with extension	Simulation period of 1984–1999			Simulation period of 2005–2016		
		t-stat	*P*-value	Ranking	t-stat	*P*-value	Ranking
1	ALPHA_BNK.rte	3.92	0.000	1	0.48	0.63	14
2	CN2.mgt	– 3.61	0.001	2	– 1.99	0.05	1
3	CH_N2.rte	3.07	0.003	3	– 0.67	0.507	11
4	CH_K2.rte	– 1.82	0.072	4	– 0.77	0.444	9
5	OV_N.hru	– 1.31	0.195	5	– 0.09	0.926	16
6	GW_REVAP.gw	– 1.18	0.242	6	– 1	0.32	6
7	RCHARG_ DP.gw	– 1.08	0.283	7	0.92	0.36	7
8	GW_DELAY.gw	0.98	0.329	8	0.61	0.543	12
9	ESCO.hru	– 0.93	0.355	9	0.12	0.906	15
10	GWQMN.gw	– 0.8	0.426	10	1.3	0.196	4
11	SURLAG.bsn	0.78	0.44	11	– 1.72	0.088	3
12	SOL_AWC.sol	– 0.73	0.47	12	– 0.71	0.481	10
13	ALPHA_BF.gw	– 0.7	0.487	13	– 1.13	0.262	5
14	SOL_BD.sol	0.66	0.512	14	0.84	0.402	8
15	REVAPMN.gw	– 0.5	0.621	15	– 1.79	0.077	2
16	SOL_K.sol	0.37	0.715	16	– 0.48	0.629	13

Source Computed by authors

Table 5.10 LULC change matrix between 1995 and 2016

LULC type (km^2)	Agricultural land	Bare ground	Built-up area	Forest	Waterbodies	Total in 2016	Net change (%) (1995–2016)
Agricultural land	4674.89 (45.14%)	32.24 (0.31%)	21.66 (0.21%)	1208.30 (11.67%)	64.48 (0.62%)	6001.58 (57.95%)	4.25
Bare ground	4.41 (0.04%)	0.09 (0.001%)	0.44 (0.004%)	0.75 (0.01%)	14.08 (0.14%)	19.77 (0.19%)	– 0.17
Urban built-up area	381.53 (3.68%)	0.50 (0.005%)	19.66 (0.19%)	23.57 (0.23%)	1.04 (0.01%)	426.31 (4.12%)	3.62
Forest	481.31 (4.65%)	4.80 (0.05%)	9.76 (0.09%)	3343.97 (32.29%)	16.31 (0.16%)	3856.15 (37.23%)	– 7.04
Waterbodies	19.11 (0.18%)	0.00 (0.00%)	0.36 (0.0003%)	8.19 (0.08%)	24.83 (0.24%)	52.50 (0.51%)	– 0.66
Total in 1995	5561.25 (53.70%)	37.63 (0.36%)	51.89 (0.50%)	4584.78 (44.27%)	120.75 (1.17%)	10,356.30 (100%)	–

Source Computed by the authors

Table 5.11 Accuracy assessment of the LULC map

LULC type	Producer's accuracy (%)		User's accuracy (%)		Overall accuracy (%)		Conditional Kappa statistics (%)		Overall, Kappa statistics (%)	
	1995	2016	1995	2016	1995	2016	1995	2016	1995	2016
Agricultural land	90.20	92.00	90.20	90.20	88.73	90.20	86.93	87.01	85.59	87.28
Bare ground	76.00	66.67	82.61	88.89			80.18	88.19		
Built-up area	82.14	87.04	82.14	92.16			79.30	89.33		
Forest	90.20	93.88	90.20	90.20			86.93	87.10		
Waterbodies	95.92	94.87	92.16	88.10			89.68	85.28		

Source Computed by the authors

Table 5.12 Results of simulated mean runoff

Change point	Simulation model	LULC for model calibration and Weather data for the model run	Simulated mean runoff (mm)
2001	Sim1 (1981–2001)	The model was calibrated to reference period LULC, and it was run using reference period weather data (1981–2001)	568.54
	Sim2 (2002–2018)	The model was calibrated to interference period LULC, and it was run using interference period weather data (2002–2018)	372.67
	Sim3 (2002–2018)	The model was calibrated to the reference period (1981–2001) LULC, but it was run using interference period (2002–2018) weather data	652.82
2008	Sim1 (1981–2008)	The model was calibrated to reference period LULC, and it was run using reference period weather data (1981–2008)	587.11
	Sim2 (2009–2018)	The model was calibrated to interference period LULC, and it was run using interference period weather data (2009–2018)	362.48
	Sim3 (2009–2018)	The model was calibrated to the reference period (1981–2008) LULC, but it was run using interference period (2009–2018) weather data	645.25

Source Computed by the authors

Table 5.13 Contribution rate of climate variability and human activities

Sl. No.	Change point	Total difference (mm)	Contribution of climate variability			Contribution rate of LULC change (%)		
			Absolute (mm)	Relative (%)	Remarks	Absolute (mm)	Relative (%)	Remarks
1	2001	– 195.87	84.28	43.03	Increasing	– 280.15	143.03	Decreasing
2	2008	– 224.64	58.13	25.88	Increasing	– 282.77	125.88	Decreasing

Source Computed by the authors

Table 5.14 Results of normality test

Sl. No.	Change point	Simulation model	Kolmogorov-Smirnov[a]			Shapiro–Wilk		
			Statistics	*p*-value	Remarks	Statistics	*p*-value	Remarks
1	2001	Sim1	0.116	0.200*	Normal	0.942	0.240	Normal
		Sim2	0.105	0.200*	Normal	0.966	0.748	Normal
		Sim3	0.127	0.200*	Normal	0.962	0.661	Normal
2	2008	Sim1	0.090	0.200*	Normal	0.969	0.564	Normal
		Sim2	0.114	0.200*	Normal	0.951	0.682	Normal
		Sim3	0.136	0.200*	Normal	0.961	0.801	Normal

* Lower bound of the true significance
[a] Lilliefors significance correction
Source Computed by the authors

increasing runoff and LULC change in decreasing runoff were 25.88% and 125.88%, respectively.

It is worth noting that the mean annual runoff was higher in the reference period of the change point in 2008 compared to the reference period of the change point in 2001. Additionally, in the interference periods, the Sim3 model of 2002–2018 exhibited a higher runoff value of 652.82 mm compared to the Sim3 model of 2009–2018, which had a runoff value of 645.25 mm.

However, the overall findings suggest that there was a negative shift (step down) in both the change points (2001 and 2008) based runoff series, and it was largely due to human intervention, while climate variability had just a reverse impact on runoff, i.e., it caused an increase in mean runoff in interference period as compared to that in the reference period.

Table 5.15 Results of t-test and estimation of effect size by Cohen's *d*

Sl. No.	Change point	Simulation model	*t*-value	*p*-value (2-tailed)	Cohen's *d*	Remarks
1	2001	Sim2 and Sim1	4.382	0.000	1.43	Significant with very large effect size
		Sim3 and Sim1	– 1.646	0.109	–	Not significant
		Sim2 and Sim3	– 6.605	0.000	2.27	Significant with huge effect size
2	2008	Sim2 and Sim1	4.078	0.000	1.50	Significant with very large effect size
		Sim3 and Sim1	– 0.967	0.340	–	Not significant
		Sim2 and Sim3	– 4.350	0.000	1.95	Significant with very large effect size

Note 1. t-test was applied, assuming equal variances between two datasets
2. Variance was compared by applying Levene's test
Source Computed by the authors

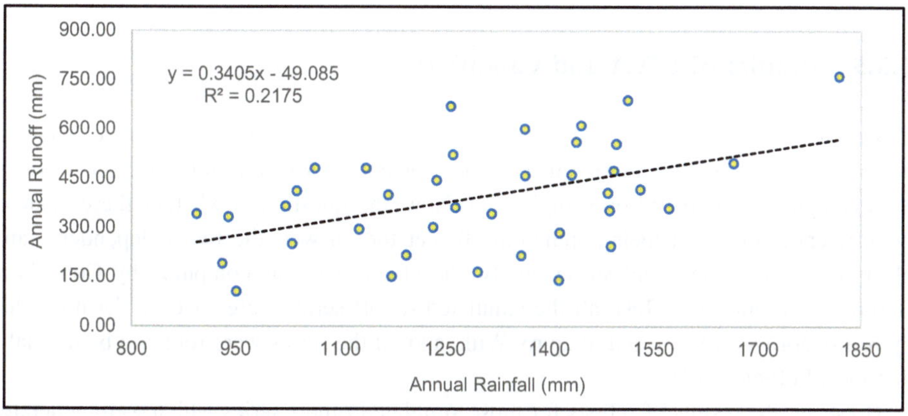

Fig. 5.7 Linear regression model fitting of rainfall versus runoff. *Source* Computed based on rainfall data from NASA Power Data Access, 2020, and CWC discharge data, 2020

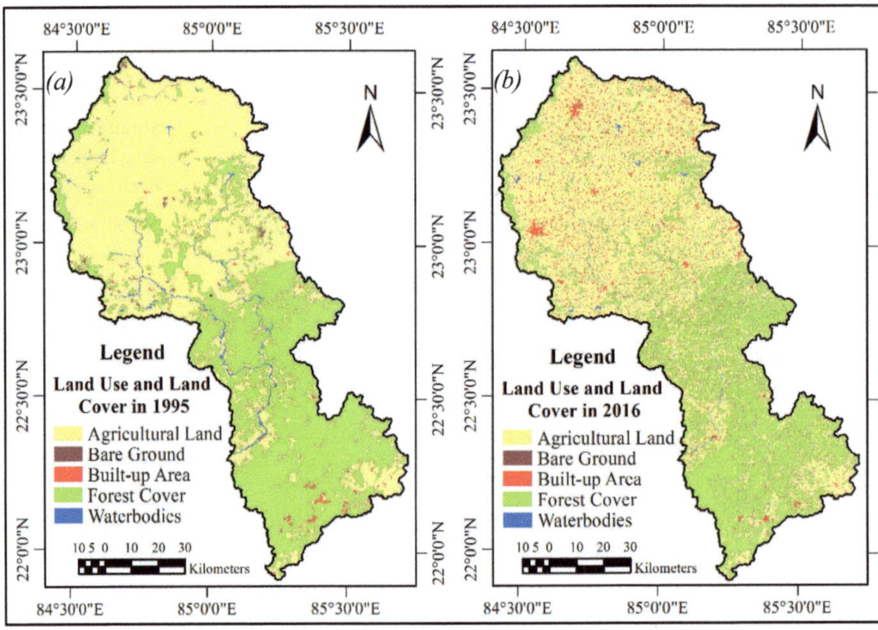

Fig. 5.8 Spatial distribution of land use and land cover in the South Koel River basin for the years **a** 1995 and **b** 2016, illustrating changes in agricultural land, forest cover, built-up areas, bare ground, and water bodies. *Source* Landsat TM data, 1995, and OLI data, 2016

5.3.5 Results of t-Test and Cohen's *d*

It is clear that the simulated mean annual runoff decreased from the respective reference periods to the respective interference periods for both 2001 and 2008 as change points. This computation is based on sample data. Now, the question is whether these changes or differences occur in their population. To get the answer, the two independent sample t-Test was applied, and subsequently, the effect size was computed by Cohen's *d*. Before performing the *t*-Test, all the simulated runoff series were checked for normality by Kolmogorov–Smirnov and Shapiro–Wilk, and all the series were found to be normally distributed (Table 5.14).

However, the results of *t*-Tests for those simulated runoff series which were generated for reference and interference period as defined by change point 2001 are as follows (Table 5.15):

(i) The effective changes in runoff caused by the combined effect of climate variability and LULC changes from the reference period (1981–2001) to the interference period (2002–2018) were found to be statistically significant. This indicates that whatever runoff difference appears, it does not appear due to sampling fluctuation; rather, it is really due to the combined effect of climate variability and human intervention.

(ii) The increase in runoff from reference to interference period which is contributed by climate change alone, is not found to be statistically insignificant. This suggests that the observed changes in runoff are happening by chance, and with a change in the sample, the result might be changed. So, the observed increase in runoff could not be attributed to climate variability alone.

(iii) However, the decrease in runoff caused by LULC change from the reference period to the interference period is again found to be statistically significant. This implies that the reduction in runoff in the interference period is not happening by chance due to sampling fluctuation; rather, LULC change is effectively causing alterations in runoff.

Now, the results of the *t*-Test for those simulated runoff series which were generated for reference and interference period as defined by change point 2008, are as follows (see Table 5.15):

(i) The effective decrease in runoff caused by the combined effect of climate variability and LULC change from the reference period (1981–2008) to the interference period (2009–2018) was found to be statistically significant. This implies that the total runoff change is not due to the chance factor but rather the combined effect of climate variability and LULC change eventually caused a reduction in runoff in the interference period as compared to that in the reference period.

(ii) The increase in runoff caused by climate variability alone was not statistically insignificant. This indicates that the observed changes in runoff is happening by chance due to sampling fluctuation, and climate variability is not the effective driving force for such alteration.

(iii) Conversely, the reduction in runoff caused by LULC change is found to be showed statistically significant. The LULC change is an effective driving force to cause alteration in runoff.

So, the test results for both the change point, i.e., 2001 and 2008, are found to be consistent with each other, although the magnitude of absolute alteration was not numerically the same. Hence, the effect size was computed by using Cohen's *d*. Its value is likely to be higher for LULC change than for climate variability for both the change points, i.e., for 2001 and 2008. The 2001 change point shows $d = 2.27$ for the LULC change, which is higher than what the 2008 change point shows for the LULC change ($d = 1.95$). The effect size for the combined influence of climate change and human activity, on the other hand, is higher ($d = 1.5$) for 2008 based interference period than 2001 based interference period ($d = 1.43$). This implies that climate variability is going to be more effective in forthcoming days in causing alteration in runoff (Table 5.15).

5.3.6 Overall Discussion

In this chapter, an attempt was made to assess the impact of climate variability and human activities (through LULC change) on the runoff. The results showed that the simulated mean annual runoff was decreased by 195.87 mm due to the combined effect of climate variability and LULC change from the period 1981–2001 to 2002–2018. This change was accelerated by 28.77 mm from 2009 onward. Hence, the effect size was higher in 2009–2018 as compared to 2001–2018, as suggested by Cohen's *d*.

On the other hand, the role of climate variability in increasing runoff was higher in 2002–2018 (84.29 mm) compared to 2009–2018 (58.13), although it was not statistically significant. The results are very much consistent with the results of trend analysis for rainfall which was discussed in Chap. 4. There was no significant trend in the rainfall series, but runoff was showing a gradual decreasing tendency. The statistically insignificant contribution of climate variability on runoff alteration, as found in the current investigation in the present chapter, substantiates the trend result and getting substantiated by trend result too.

Now, the question is which one is then the main driving force behind the significant decrease in runoff as found in runoff trend analysis. The answer lies in the relative contribution of LULC change on runoff alteration. The statistically significant contribution

Fig. 5.9 Field photograph showing agricultural land, cropping practices, and the gauge station of South Koel River in the study area, captured using a geotagged mobile camera during ground-truth verification

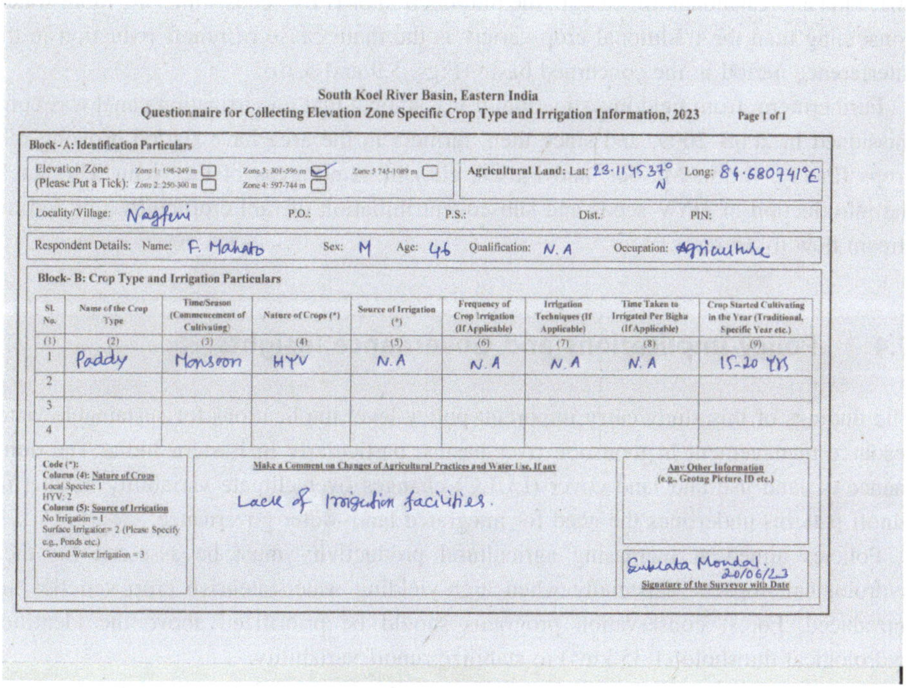
South Koel River Basin, Eastern India

Questionnaire for Collecting Elevation Zone Specific Crop Type and Irrigation Information, 2023 Page 1 of 1

Block - A: Identification Particulars

Elevation Zone (Please Put a Tick): Zone 1: 198-249 m ☐ Zone 2: 250-300 m ☐ Zone 3: 301-596 m ☑ Zone 4: 597-744 m ☐ Zone 5 745-1089 m ☐

Agricultural Land: Lat: 23.114537° N Long: 84.680741°E

Locality/Village: Nagfeni P.O.: P.S.: Dist.: PIN:

Respondent Details: Name: F. Mahato Sex: M Age: 46 Qualification: N.A Occupation: Agriculture

Block- B: Crop Type and Irrigation Particulars

Sl. No.	Name of the Crop Type	Time/Season (Commencement of Cultivating)	Nature of Crops (*)	Source of Irrigation (*)	Frequency of Crop Irrigation (If Applicable)	Irrigation Techniques (If Applicable)	Time Taken to Irrigated Per Bigha (If Applicable)	Crop Started Cultivating in the Year (Traditional, Specific Year etc.)
(1)	(2)	(3)	(4)	(5)	(6)	(7)	(8)	(9)
1	Paddy	Monsoon	HYV	N.A	N.A	N.A	N.A	15-20 Yrs
2								
3								
4								

Code (*):
Column (4): Nature of Crops
Local Species: 1
HYV: 2
Column (5): Source of Irrigation
No Irrigation = 1
Surface Irrigation= 2 (Please Specify e.g., Ponds etc.)
Ground Water Irrigation = 3

Make a Comment on Changes of Agricultural Practices and Water Use, If any

Lack of Irrigation facilities.

Any Other Information (e.g., Geotag Picture ID etc.)

Subrata Mondal. 20/06/23
Signature of the Surveyor with Date

Fig. 5.10 Sample questionnaires used in the field survey

of LULC change to the reduction in runoff provides evidence that LULC change is the main causal factor for the reduction in runoff in South Koel River Basin.

The LULC analysis shows that the most dominant LULC in the South Koel River basin is agricultural, and its areal coverage increased from 53.70% in 1995 to 57.95% in 2016 and occurred at the cost of forest land which decreased from 44.27% in 1995 to 37.23% in 2016. The conversion of forest to agricultural land is usually supposed to increase the runoff, but instead of increasing, the runoff decreased in the interference period as compared to that in the reference period. This clearly indicates that not merely agricultural land but rather agricultural technology has an important role in this runoff alteration.

An extensive field survey was done by the researcher to get insight into agricultural technology in the study area. Paddy is the principal crop all over the study area, and it is usually cultivated once a year during monsoon. Since the last 15–20 years, the farmers of the concerned area have been extensively using high-yielding variety (HYV) paddy and many other crops, all of which are more water-consuming crops. Some notable HYV paddy seeds, as reported by local farmers, are CR Dhan 202, CR Dhan 204, CR Dhan

201, Puja and Chandankitli, etc. So, the introduction of HYV seeds which are more water-consuming than the traditional crop variety is the main cause of runoff reduction in the interference period in the concerned basin (Figs. 5.9 and 5.10).

Furthermore, from field investigation, it was known that one irrigation canal was commissioned in 2008–2009, and since then, farmers in the area have started growing rabi crops (Paddy, Gram, Masoor, and Pulses, etc.) in some areas of the basin. Ultimately the introduction of HYV seeds and subsequent initiation of rabi crop cultivation led the stream flow to decrease.

5.4 Policy Implications and Governance Insights

The findings of this study carry important policy-level implications for sustainable water resource management in monsoon river basins, particularly in Eastern India. The dominance of land use and land cover (LULC) changes over climate variability in altering runoff patterns underlines the need for integrated land–water governance.

Policies aimed at increasing agricultural productivity must be assessed for their hydrological impacts, especially when high-yielding water-intensive crop varieties are introduced. Forest conservation programs should be prioritized above the identified hydrological threshold (1.35 km^2) to stabilize runoff variability.

Moreover, decentralized watershed management, adoption of climate-smart agricultural practices, and region-specific land zoning regulations could help reduce human-induced hydrologic disruption. The results suggest that runoff can be partially restored if human interference is minimized and vegetation cover is actively maintained, especially in upper and middle sub-basins.

These findings can guide the Jharkhand and Odisha state governments, Central Water Commission, and policy think-tanks such as NITI Aayog to refine water policy, encourage stakeholder collaboration, and support data-driven land use planning.

5.4.1 Limitations of the SWAT Model and Hydrological Uncertainties

Although the SWAT model has proven highly effective for simulating runoff responses under various scenarios, certain limitations are inherent. The discretization into Hydrologic Response Units (HRUs) may lead to spatial generalization that omits small-scale processes like local depressions or lateral water transfer between units. Additionally, input data such as soil properties, land cover classification, and rainfall distribution—especially in data-scarce tropical regions—introduce uncertainties. Model calibration and validation mitigate these limitations, but the results must still be interpreted in the context of these constraints. Future studies could benefit from higher-resolution data, inclusion of evapotranspiration dynamics, and uncertainty analysis using ensemble runs or Bayesian techniques.

5.5 Findings in Brief

The main objective that has been incorporated in this chapter was to separate out the contribution of climate variability and LULC change in the alteration of runoff. Following the methodology, six SWAT simulation models were formulated, three for 2001-based reference and interference periods and another three for 2008-based reference and interference periods. To fulfil the objective all the model outputs were compared by applying simple arithmetic. The findings are summarized below:

(i) Calibration and validation of all six simulation models, as evaluated by NSE and R^2, were satisfactory. NSE and R^2 for calibration range from 0.77 to 0.83 and 0.77 to 0.86, respectively. Furthermore, the model's performance was reasonably accurate, as suggested by the NSE (0.77–0.91) and R^2 (0.79–0.93) values for validation.

(ii) Global sensitivity analysis showed that ALPHA_BNK, CN2, and CH_N2 were the most sensitive parameters, respectively, in influencing the runoff patterns for the reference period during 1981–2001, while CN2 was the most sensitive parameter for the interference period during 2002–2018.

(iii) Climate variability causes an increase in runoff in the interference period as compared to that in the reference period, whereas LULC change leads to a decrease in runoff for the same period. The contribution rate of climate variability to increase the runoff was 43.03% and 25.88% in 2002–2018 and 2009–2018, respectively, while the contribution rate of human activities (through LULC change) in decreasing runoff was 143.03% and 125.88% in 2002–2018 and 2009–2018 respectively.

(iv) The increase in runoff due to climate variability is not statistically significant, as suggested by t-Test results, whereas the role of LULC change in decreasing the runoff is found to be significant. This implies that LULC change is the effective driving force for the reduction in runoff in the recent past (2002–2018). The effect size of changing LULC upon the runoff reduction, as evaluated by Cohen's d value, is likely to be greater in the period of 2009–2018 ($d = 1.95$) compared to 2002–2018 ($d = 1.43$) as suggested by Cohen's d.

(v) Agricultural land is the most dominant LULC class in the basin. It increased from 53.70% in 1995 to 57.59% in 2016, whereas the forest cover decreased from 44.27% in 1995 to 37.23% in 2016. This conversion of forest to agricultural land is supposed to increase the runoff, but, in the basin, instead of increasing runoff, it has decreased since 2002 onward. The ground truth verification substantiates that it is the new agricultural technology that incorporates more water-consuming HYV seeds that is acting as the main driving force for reducing the runoff.

5.6 Concluding Remarks

In this chapter, the impact of climate variability and human activities on runoff alteration was assessed for the South Koel River basin for the period of 1981–2018. It is observed that LULC change is effectively causing the reduction in runoff since 2002 onward, as suggested by SWAT model analysis and subsequent hypothesis test by means of t-Test. Contrary to this, climate variability causes an increase in runoff, but its role is not statistically significant. The findings of the numeric analysis and ground truth verification provide sufficient evidence to conclude that it is the HYV-based agricultural technology which is playing an effective role behind the runoff alteration in the basin. Therefore, the analyses presented herein satisfactorily address the second research objective.

Runoff Sensitivity to LULC Change 6

6.1 Prelude

The third objective of the study is *to capture the runoff sensitivity to changes in individual land use and land cover (LULC) classes with special reference to forest cover, agricultural land use, and urban built-up area in the South Koel River Basin.* Specifically, the analysis examines how changes in the areal coverage of each class alter runoff dynamics, i.e., the role of each factor in controlling the runoff. Each factor's role is assessed while controlling for the others, with the goal of identifying the threshold for that factor beyond which changes significantly affect runoff.

Understanding runoff sensitivity to these land use categories is essential for determining the thresholds of land conversion that trigger substantial impacts. This approach provides a novel quantitative assessment of LULC patch-size effects on runoff behavior and offers valuable insights for sustainable land and water management in tropical river basins. The findings are expected to inform region-specific strategies for integrated water governance.

6.2 Database and Methods

6.2.1 Data Generation

A one-way ANCOVA model was run for achieving the third research objective. The methodology was explained in detail in Sect. 3.3. Here, how the input database for the ANCOVA model was generated by applying SWAT simulation is described below.

S. Mondal and R. K. Paul, *Climate Variability and Human Interference*, Deltas of the World, https://doi.org/10.1007/978-3-032-13257-4_6

The SWAT model was set up, calibrated, and validated using the DEM, soil, and LULC datasets, and then the model was run by the weather datasets to simulate the annual runoff for the period of 1981–2001.

In the model setup, twenty-seven subbasins were generated, and all the subbasins had further subdivided the subbasins into 1036 homogeneous units called Hydrologic Response Units (HRUs). Each HRU is a unique areal unit having uniformity in LULC, soil type, and slope characteristics, i.e., one HRU is occupied by only one LULC class, one soil type, and one slope class. If one HRU is occupied by forest, it does not have other LULC classes, e.g., agriculture or built-up, etc. Here, all these parameters, i.e., LULC, soil, and slope, were included to form HRUs if their areal coverage exceeds the threshold value of 1%. Ultimately the HRUs that were produced were of different sizes.

After the simulation, the output hydrologic variables of rainfall, runoff, etc., were generated for each HRU on a monthly basis for the period 1981–2001. The monthly rainfall and runoff output of the period 1981–2001 were transformed into annual average value, i.e., one rainfall value and one runoff value for each HRU was generated.

Out of the total HRUs (1036), 544 were occupied by only forest cover, 364 were occupied by agricultural land, 36 were occupied by built-up areas, and the remaining were occupied by other LULC classes.. For each type of HRU, a separate data table was generated, i.e., one data table for forest cover HRUs, one for agriculture land HRUs, and one for urban built-up area HRUs. Each data table contained three columns, one for HRU type, one for rainfall, and one for runoff. The HRUs of each data table was classified into four groups based on the area of HRU using the quantile classification method (Table 6.1).

Table 6.1 Categorization of independent variables and their respective HRUs

Variable's name	Number of classes	Category	Number of HRUs
Forest cover (km^2)	4	<1.35	137
		1.35–3.44	138
		3.44–9.14	133
		>9.14	136
Agricultural land (km^2)	4	<1.30	92
		1.30–2.86	90
		2.86–10.50	92
		>10.50	90
Urban built-up area (km^2)	4	<0.74	10
		0.74–1.05	8
		1.05–1.51	8
		>1.51	10

Source Computed by the authors

Now, the input database for the ANCOVA model is ready in which runoff is the response variable, HRU class is the explanatory variable which is a categorical one with four HRU size classes, and rainfall is the covariate. Here, three ANCOVA models were applied separately for each data table, i.e., one ANCOVA was applied using the input database on forest HRU, one by using the agriculture HRU database, and one for the built-up HRU database.

It is worth noting that one-way ANCOVA was applied to judge whether the mean runoff varies across the HRU class so that the threshold value of HRU size class above or below which the runoff shows significant difference can be determined, and obviously, it was done after controlling the rainfall. In the present study, in addition to one-way ANCOVA, one-way ANOVA was also applied to see whether both models produced any large-scale difference in output. Logically ANCOVA is the best-suited tool for data analysis in order to achieve the third research objective, but ANOVA was also applied only to determine the runoff response to HRU size class without controlling the rainfall. Hence, the ultimate evaluation was done based on the ANCOVA model output.

6.2.2 Method of Analysis

To assess the runoff sensitivity to changing areal coverage of three specific individual LULC classes: forest cover, agriculture, and built-up area, the whole basin area was divided into a manageable number of homogeneous units technically called *Hydrologic Response Unit* (HRU) in such a way that each HRU contains only a single LULC class, i.e., either forest or agriculture or built-up, etc. Then the SWAT model was run for the period 1981–2001, and the average annual runoff was computed for each of the HRU(s) belonging to only LULC classes forest, agriculture, and built-up. The runoff values of each individual HRU belonging to the forest, agriculture, and built-up were further classified into four sub-categories based on the areal coverage of corresponding HRU(s) by using the quantile method and then three separate ANCOVA models with rainfall as a covariate were run for three respective LULC classes to analyse the sensitivity of runoff to changing areal coverage of respective HRU(s) belonging to corresponding LULC class (Fig. 6.1).

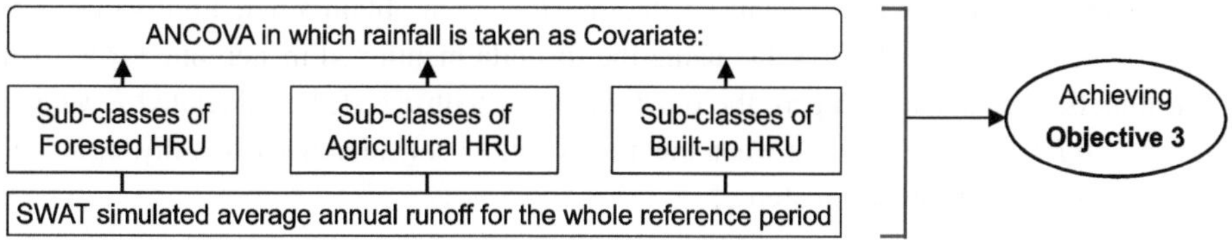

Fig. 6.1 Methodological design that was adopted for achieving third research objective

As mentioned earlier, logically, ANCOVA is the best-suited tool for achieving the third research objective, but ANOVA was also applied only to determine the runoff response to HRU size class without controlling the rainfall. The ultimate evaluation was done based on the ANCOVA model output. Here, statistical tools, one-way ANOVA and one-way ANCOVA, are described below.

(a) *One-Way ANOVA*

One-Way ANOVA is a statistical method used to judge whether the mean runoff values vary across each category of the respective HRU. Three separate ANOVA models were applied, one for forested HRU, one for agriculture HRU, and one for urban built-up HRU. In each model, HRU was the fixed factor, with four groups made based on the size of HRU, and the runoff was the response variable. This method is unable to take control of the effect of rainfall on runoff variation.

Computation was done using SPSS v.17.0. Model involves two steps: first, classification of the total sum of square deviation (SST) of the response variable (runoff) into the sum of squares between the group (SSB) and the sum of squares within the group (SSW) and second, application of F test to judge whether mean SSB is larger than mean SSW. SSB is the treatment effect, i.e., explained variance, while SSW is the error, i.e., unexplained variance. If mean SSB is proved to be greater than SSW at a 5% significance level; the mean runoff is assumed to differ at least in between two size classes of HRU.

If the F-test shows statistically significant differences, it is important to determine which specific groups exhibit these differences, and it was done by applying a post hoc test. Tukey's HSD was applied for the condition of equal variance in runoff across the size class of HRU; otherwise, the Games Howell method was applied for pair-wise comparison of mean runoff. The equality of variance was tested by using Levene's test.

The procedure outlined by Petz (1981) involves the following steps:

Step One: Calculate the F values for each pair of arithmetic means using the formula:

$$F = \frac{(M_{ai} - M_b)^2}{MS_{wg}(N_a + N_b) \div N_a N_b} \tag{6.1}$$

Step Two: Determine the critical F-value for the desired significance level using the degrees of freedom ($(k - 1)$ and $(N - 1)$) and refer to an F table.

Step Three: Multiply the critical F value by $(k - 1)$ to obtain a new limit value ($F\prime$).

Step Four: Calculate the F value using the formula mentioned in step one for all pairs of arithmetic means and compare the obtained value with $F\prime$. If the calculated F value is higher than $F\prime$, the difference between the means can be considered statistically significant at the significance level set in step 2.

(b) *One-Way ANCOVA*

One-Way ANCOVA was also used to show whether the mean runoff values vary across the HRU size class after controlling the rainfall for the respective HRU class. Three separate ANCOVA models were run, one for forest HRU, one for agriculture HRU, and one for built-up HRU. In each type of model, HRU was the fixed factor having four groups made based on the area of respective HRU, and runoff was the response variable, whereas rainfall was the covariate. Computation was done using SPSS v.17.0.

A typical ANCOVA model, assuming that the slopes and intercepts for the two groups are potentially different, can be represented as follows (Helsel and Hirsch 1995)

$$y = (\beta_0 + \beta_2 Z) + (\beta_1 + \beta_3 Z)x + \varepsilon \tag{6.2}$$

In this equation, Z is a binary variable that takes the value 0 or 1 depending on the group to which x and y belong. β_0 and $\beta_0 + \beta_2$ represent the intercepts during the calibration and treatment periods, respectively. β_1 and $\beta_1 + \beta_3$ represent the slopes during the calibration and treatment periods, respectively. If β_2 is non-zero and β_3 is zero, the regression lines produced by the equation will be parallel. However, if both β_2 and β_3 are non-zero, the regression lines will be distinct for the two groups. After applying ANCOVA post hoc test was done to determine the pair-wise comparison in the same way as it was done for ANOVA.

6.3 Results and Discussion

6.3.1 Runoff Sensitivity to Forest Cover

The results of ANOVA show that there is a significant difference in mean runoff between at least two groups of forested HRU at a 5% significance level (Table 6.2). This implies that runoff is sensitive to areal coverage of forest cover, i.e., with a change in forest cover size, runoff changes. The partial eta squared value indicates that the 1.9% variation in runoff is explained by the difference in the size class of forested HRU. But how much changes in areal coverage of forest cover create a significant difference in runoff could be known from the pair-wise comparison of the mean runoff of one size class of forested HRU to the mean runoff of another size class of forested HRU what has been done in posthoc analysis.

The results of the post hoc test show that the surface runoff was significantly increased from the forest area of <1.35 to 1.35–3.44 km^2 but beyond the range of 1.35–3.44 km^2; further increases in forest area did not cause the alteration of surface runoff. There was no significant variation in surface runoff when the forest area increased from 1.35–3.44 to 3.44–9.14 km^2. Only when the forest area exceeded 9.14 km^2 the surface runoff again showed a significant difference compared to the range of 1.35–3.44 km^2 (Table 6.3).

Table 6.2 Results of ANOVA for forest cover

Sources of variance	Sum of squares	*df*	Mean square	F	Sig	Partial eta squared	Remarks
Between groups	47,556.21	3	15,852.07	3.518	0.015	0.019	Significant
Within groups	2,433,231.7	540	4505.98				
Total	2,480,787.9	543					

Dependent variable: surface runoff (mm)
Factor: forest cover (km^2)
Source Computed by the authors

It can be mentioned that increasing the forest area beyond 1.35 km^2 led to changes in surface runoff without controlling the effect of rainfall and other factors. The runoff further changes if the forest area again goes beyond 9.14 km^2. Therefore, the threshold value for forest cover in influencing surface runoff was determined to be 1.35 km^2. So, the relationship between forest cover and runoff is not linear one.

When the influence of rainfall was eliminated, the mean annual surface runoff still exhibited significant variations between different groups or categories, as indicated by the results of ANCOVA at a significance level of 0.05 (Table 6.4). After controlling rainfall, forest cover is again found to explain the 1.9% of the total variance in surface runoff as suggested by partial eta squared suggested the value of 0.019. So, the rainfall variation within forest cover does not cause alteration in runoff generation; rather, the size of forest cover effectively affects the runoff generation mechanism.

The pairwise comparisons revealed that the mean surface runoff of forested HRU size class <1.35 km^2 significantly differs from that in the size class 1.35–3.44, 3.44–9.14, and >9.14 km^2. When the forest area was increased from less than 1.35 to 1.35–3.44 km^2, there was a significant difference in surface runoff. However, there was no significant difference in surface runoff when comparing the runoff in forest areas between 1.35–3.44 and 3.44–9.14 km^2. Beyond 9.14 km^2, increasing the forest area again resulted in an insignificant difference in surface runoff. Furthermore, if the forest area was increased from 3.44 to 9.14 km^2, the surface runoff did not show any changes. It can be stated that increasing the forest area beyond 1.35 km^2 leads to changes in surface runoff after removing the influence of rainfall. Therefore, the threshold value for forest cover in influencing surface runoff was determined to be 1.35 km^2 (Table 6.5).

Table 6.3 Results of post hoc test of ANOVA for forest cover

Test	(I) Forest cover	(J) Forest cover	Mean difference (I–J)	Std. error	Sig	95% confidence interval		Remarks
						Lower bound	Upper bound	
Games-Howell	<1.35	1.35–3.44	−23.28*	8.41	0.03	−45.02	−1.55	**Significant**
		3.44–9.14	−21.79	8.7	0.061	−44.27	0.7	NOT
		>9.14	−18.32	8.23	0.119	−39.59	2.96	NOT
	1.35–3.44	<1.35	23.28*	8.41	0.03	1.55	45.02	**Significant**
		3.44–9.14	1.5	8.06	0.998	−19.33	22.32	NOT
		>9.14	4.97	7.55	0.912	−14.54	24.48	**Significant**
	3.44–9.14	<1.35	21.79	8.7	0.061	−0.7	44.27	NOT
		1.35–3.44	−1.5	8.06	0.998	−22.32	19.33	NOT
		>9.14	3.47	7.87	0.971	−16.87	23.81	NOT
	>9.14	<1.35	18.32	8.23	0.119	−2.96	39.59	NOT
		1.35–3.44	−4.97	7.55	0.912	−24.48	14.54	NOT
		3.44–9.14	−3.47	7.87	0.971	−23.81	16.87	NOT

Dependent variable: surface runoff (mm)
Factor: forest cover (km^2)
Bold indicates statistical significance which refer to differences in surface runoff associated with variations in land-use and land-cover size classes are unlikely to have occurred by chance alone; results with *p*-values below 0.05 are considered significant
Source Computed by the authors

Table 6.4 Results of ANCOVA for forest cover

Source of variance	Type III sum of squares	*df*	Mean square	F	Sig	Partial eta squared	Remarks
Forest cover (km^2)	47,264.42	3	15,754.81	3.494	0.016	0.019	**Significant**
Rainfall (mm)	3049.97	1	3049.97	0.676	0.411	0.001	NOT
Error	2,430,181.7	539	4508.69				
Corrected total	2,480,787.9	543					

Dependent variable: surface runoff (mm)
Factor: forest cover (km^2)
Covariate: rainfall (mm)
Bold indicates statistical significance which refer to differences in surface runoff associated with variations in land-use and land-cover size classes are unlikely to have occurred by chance alone; results with *p*-values below 0.05 are considered significant
Source Computed by the authors

6.3.2 Runoff Sensitivity to Agricultural Land

The findings revealed a noteworthy distinction in the average annual surface runoff among different size classes of agricultural HRU. The between-group variation in surface runoff is significantly larger than the withing group runoff variation at a 5% significance level. Furthermore, the size of the agricultural land was found to explain ~3.1% of the overall variation observed in the surface runoff (Table 6.6).

The findings of pair-wise comparison in runoff of different size groups of agricultural HRU indicate that there is a significant difference in surface runoff when comparing agricultural land sizes below 1.30 km^2 to sizes between 1.30 and 2.86 km^2. Similarly, there is a significant difference when comparing agricultural land sizes between 2.86 and 10.50 km^2 to sizes <10.50 km^2. However, no significant differences were observed in comparisons between other different size categories within agricultural land. These results suggest that agricultural land size is not linearly related to runoff. Based on these results, it can be stated that increasing the agricultural land area beyond 1.30 km^2 leads to changes in surface runoff. Therefore, the threshold value for agricultural land size in influencing surface runoff was determined to be 1.30 km^2 (Table 6.7).

Even when the influence of rainfall was removed, significant variations in mean annual surface runoff were still observed among different sizes of agricultural land, as indicated by the results of ANOVA, in which rainfall was taken as a covariate. The inclusion of rainfall as a covariate had a significant influence on rainfall variation. The effect sizes (partial eta squared) indicated that agricultural land explains ~2.3% of the variance in

Table 6.5 Results of pairwise comparisons of ANCOVA for forest cover

(I) Forest cover	(J) Forest cover	Mean difference (I–J)	Std. error	Sig.[a]	95% confidence interval for difference[a]		Remarks
					Lower bound	Upper bound	
<1.35	1.35–3.44	−23.34*	8.1	0.004	−39.25	−7.43	**Significant**
	3.44–9.14	−21.82*	8.17	0.008	−37.87	−5.76	**Significant**
	>9.14	−17.58*	8.18	0.032	−33.64	−1.51	**Significant**
1.35–3.44	<1.35	23.34*	8.1	0.004	7.43	39.25	**Significant**
	3.44–9.14	1.52	8.16	0.852	−14.51	17.55	NOT
	>9.14	5.76	8.17	0.481	−10.29	21.81	NOT
3.44–9.14	<1.35	21.82*	8.17	0.008	5.76	37.87	**Significant**
	1.35–3.44	−1.52	8.16	0.852	−17.55	14.51	NOT
	>9.14	4.24	8.24	0.607	−11.95	20.43	NOT
>9.14	<1.35	17.58*	8.18	0.032	1.51	33.64	**Significant**
	1.35–3.44	−5.76	8.17	0.481	−21.81	10.29	NOT
	3.44–9.14	−4.24	8.24	0.607	−20.43	11.95	NOT

Based on estimated marginal means
*The mean difference is significant at the 0.05 level
[a]Adjustment for multiple comparisons: least significant difference (equivalent to no adjustments)
Dependent variable: surface runoff (mm)
Factor: forest cover (km^2)
Covariate: rainfall (mm)
Bold indicates statistical significance which refer to differences in surface runoff associated with variations in land-use and land-cover size classes are unlikely to have occurred by chance alone; results with *p*-values below 0.05 are considered significant
Source Computed by the authors

Table 6.6 Results of ANOVA for agricultural land

Sources of variance	Sum of squares	*df*	Mean square	F	Sig	Partial eta squared	Remarks
Between groups	78,148.71	3	26,049.57	3.865	0.01	0.031	**Significant**
Within groups	2,426,508	360	6740.3				
Total	2,504,656.7	363					

Dependent variable: surface runoff (mm)
Factor: agricultural land (km^2)
Bold indicates statistical significance which refer to differences in surface runoff associated with variations in land-use and land-cover size classes are unlikely to have occurred by chance alone; results with *p*-values below 0.05 are considered significant
Source Computed by the authors

Table 6.7 Results of post hoc test of ANOVA for agricultural land

Test	(I) Agricultural land	(J) Agricultural land	Mean difference (I–J)	Std. error	Sig	95% confidence interval		Remarks
						Lower bound	Upper bound	
Games-Howell	<1.30	1.30–2.86	−37.16*	12.7	0.02	−70.09	−4.23	**Significant**
		2.86–10.50	−24.54	12.72	0.22	−57.53	8.45	NOT
		>10.50	−6.6	12.57	0.953	−39.19	26	NOT
	1.30–2.86	<1.30	37.16*	12.7	0.02	4.23	70.09	**Significant**
		2.86–10.50	12.62	11.74	0.705	−17.81	43.05	NOT
		>10.50	30.56*	11.57	0.044	0.57	60.56	**Significant**
	2.86–10.50	<1.30	24.54	12.72	0.22	−8.45	57.53	NOT
		1.30–2.86	−12.62	11.74	0.705	−43.05	17.81	NOT
		>10.50	17.94	11.59	0.411	−12.12	48.01	NOT
	>10.50	<1.30	6.6	12.57	0.953	−26	39.19	NOT
		1.30–2.86	−30.56*	11.57	0.044	−60.56	−0.57	**Significant**
		2.86–10.50	−17.94	11.59	0.411	−48.01	12.12	NOT

*The mean difference is significant at the 0.05 level

N.B: variance was compared by applying Levene's test (variance significant)

Dependent variable: surface runoff (mm)

Factor: agricultural land (km^2)

Bold indicates statistical significance which refer to differences in surface runoff associated with variations in land-use and land-cover size classes are unlikely to have occurred by chance alone; results with *p*-values below 0.05 are considered significant

Source Computed by the authors

Table 6.8 Results of ANCOVA for agricultural land

Source of variance	Type III sum of squares	*df*	Mean square	F	Sig	Partial eta squared	Remarks
Agricultural land (km^2)	55,718.231	3	18,572.74	2.878	0.036	0.023	**Significant**
Rainfall (mm)	109,398.85	1	109,398.85	16.95	0	0.045	**Significant**
Error	2,317,109.1	359	6454.34				
Corrected total	2,504,656.7	363					

Dependent variable: surface runoff (mm)
Factor: agricultural land (km^2)
Covariate: rainfall (mm)
Bold indicates statistical significance which refer to differences in surface runoff associated with variations in land-use and land-cover size classes are unlikely to have occurred by chance alone; results with *p*-values below 0.05 are considered significant
Source Computed by the authors

surface runoff, while rainfall explains about 4.5% of the variance. Both variables interactively affect the observed runoff, but when considering only agricultural land size without controlling the rainfall effect, its explanatory power decreases to 2.3% as compared to 3.1% when rainfall was controlled. However, both agricultural land and rainfall significantly predict the outcome, and their relationships play a meaningful role in explaining the observed variability in surface runoff (Table 6.8).

The results of the pairwise comparisons indicated that when considering the impact of agricultural land on mean annual surface runoff, after removing the effects of rainfall, there was still a significant difference between agricultural land sizes below 1.30 km^2 and sizes between 1.30 and 2.86 km^2, as indicated by a posthoc analysis. However, no significant differences were found when comparing other different size categories within agricultural land. From these findings, it can be mentioned that increasing agricultural land beyond 1.30 km^2 leads to changes in surface runoff, independent of rainfall. Therefore, the threshold value for agricultural land in influencing surface runoff was determined to be 1.35 km^2 (Table 6.9).

6.3.3 Runoff Sensitivity to Urban Built-Up Area

The results shows that the runoff is not sensitive at all to size differences in urban built-up areas. The findings revealed that the between-group differences in the outcome were not statistically significant, as indicated by an F Test at a 5% significance level, although

Table 6.9 Results of pairwise comparisons of ANCOVA for agricultural land

(I) Agricultural land	(J) Agricultural land	Mean difference (I–J)	Std. error	Sig.[a]	95% confidence interval for difference[a]		Remarks
					Lower bound	Upper bound	
<1.30	1.30–2.86	−33.42*	11.95	0.005	−56.91	−9.93	**Significant**
	2.86–10.50	−23.02	11.85	0.053	−46.33	0.29	NOT
	>10.50	−12.04	11.98	0.316	−35.61	11.53	NOT
1.30–2.86	<1.30	33.42*	11.95	0.005	9.93	56.91	**Significant**
	2.86–10.50	10.4	11.92	0.384	−13.05	33.85	NOT
	>10.50	21.38	12.18	0.08	−2.58	45.34	NOT
2.86–10.50	<1.30	23.02	11.85	0.053	−0.29	46.33	NOT
	1.30–2.86	−10.4	11.92	0.384	−33.85	13.05	NOT
	>10.50	10.98	12.03	0.362	−12.68	34.64	NOT
>10.50	<1.30	12.04	11.98	0.316	−11.53	35.61	NOT
	1.30–2.86	−21.38	12.18	0.08	−45.34	2.58	NOT
	2.86–10.50	−10.98	12.03	0.362	−34.64	12.68	NOT

Based on estimated marginal means
*The mean difference is significant at the 0.05 level
[a]Adjustment for multiple comparisons: least significant difference (equivalent to no adjustments)
Dependent variable: surface runoff (mm)
Factor: agricultural land (km^2)
Covariate: rainfall (mm)
Bold indicates statistical significance which refer to differences in surface runoff associated with variations in land-use and land-cover size classes are unlikely to have occurred by chance alone; results with *p*-values below 0.05 are considered significant
Source Computed by the authors

Table 6.10 Results of ANOVA for the urban built-up area

Sources of variance	Sum of squares	*df*	Mean square	F	Sig	Partial eta squared	Remarks
Between groups	15,734.05	3	5244.68	2.841	0.053	0.21	**Insignificant**
Within groups	59,074.05	32	1846.06				
Total	74,808.1	35					

Dependent variable: surface runoff (mm)
Factor: urban built-up (km^2)
Bold indicates statistical significance which refer to differences in surface runoff associated with variations in land-use and land-cover size classes are unlikely to have occurred by chance alone; results with *p*-values below 0.05 are considered significant
Source Computed by the authors

the effect size represented by the partial eta squared value of 0.21, showed that ~21% of the total variance in the outcome could be attributed to the between-group differences. The post hoc analysis is not relevant here since the ANOVA shows insignificant results (Table 6.10).

The findings shows that the results of ANCOVA in which rainfall was the control variable, i.e., the effects of rainfall on surface runoff generation were eliminated before analysing the variance in the runoff. However, as expected, the differences between the urban built-up areas in relation to the surface runoff were not statistically significant at a 5% significance level, as indicated by an F-value of 2.789 and a *p*-value of 0.057. The effect size, represented by the partial eta squared value of 0.213, suggested that ~21.3% of the total variance can be attributed to the differences in urban built-up areas (Table 6.11).

However, the effect size is relevant here since the differences in runoff across different size groups of urban built-up HRU. Similarly, post-hoc analysis for pair-wise comparison is not also relevant in the present case.

6.4 Findings in Brief

The findings of the chapter are summarized below.

(i) The mean surface runoff varies significantly with variations in the size of forested land with or without controlling the effect of rainfall. A similar result of variance analysis with or without controlling the rainfall clearly implies that size variation in forest land is more effective than the rainfall variation within forested areas in causing variation in the runoff. This is substantiated by partial eta square value, which states that the size variation in forest land explains 1.9% of total runoff variance

Table 6.11 Results of ANCOVA for the urban built-up area

Source	Type III sum of squares	*df*	Mean square	F	Sig	Partial eta squared	Remarks
Urban built-up (km^2)	15,710.71	3	5236.9	2.789	0.057	0.213	NOT
Rainfall (mm)	864.95	1	864.95	0.461	0.502	0.015	NOT
Error	58,209.09	31	1877.71				
Corrected total	74,808.1	35					

Dependent variable: surface runoff (mm)
Factor: urban built-up (km^2)
Covariate: rainfall (mm)
Source Computed by the authors

with or without controlling rainfall, and rainfall alone explains 0.1%. The relationship between the size of forest land and the runoff variation is not linear since runoff does not vary across all the class groups of forested HRU. However, the threshold value of forest land beyond which runoff starts changing is 1.32 km^2. It is important to note that forest occupies 37.23% of the total basin area as of 2016, and so it is expected to have a noticeable influence on the runoff generation mechanism in the concerned basin.

(ii) Agricultural land size has a significant influence on surface runoff with or without controlling rainfall. The threshold value for agricultural land size in influencing surface runoff is determined to be 1.30 km^2. The relationship is not again linear one since runoff does not change significantly across all the size classes of agricultural HRU. However, agricultural land explains approximately 2.3% of the variance in surface runoff, but this increases to 3.1% when the rainfall effect is controlled. It is important to note that agricultural land is the predominant LULC class and covers a substantial portion, 57.95%, of the total basin area as of 2016. Thus, it is expected to have a significant impact on runoff variation.

(iii) The impact of changing the size of urban built-up areas on surface runoff did not yield significant results with or without controlling the rainfall effect. It is because the urban built-up area occupies only 4.12% of the total basin area as of 2016. In 1995 its percentage share was only 0.50% in 1995. And urban area is scattered in distribution within the basin. Hence, the built-up area cannot effectively alter the runoff generation mechanism in the concerned basin, but it is supposed to be a very influential factor for alteration in runoff in the watershed, which has built up as the dominant LULC class.

6.5 Concluding Remarks

The sensitivity analysis of runoff has provided valuable insights into the impacts of three key LULC classes, namely forest cover, agricultural land, and urban built-up area on surface runoff within the study area. The findings suggested that increasing forest area beyond the threshold value of 1.35 km^2 can lead to changes in surface runoff, indicating the importance of maintaining and managing forests to mitigate runoff and preserve water resources. Agricultural land size also played a significant role, with larger agricultural areas exhibiting notable effects on surface runoff. These results emphasized the need for appropriate land management practices in agricultural areas to mitigate runoff and protect water quality. The impact of urban built-up areas on surface runoff was inconclusive due to the small spatial coverage and scattered distribution in the study area. However, the present analysis focused on the influences of forest area, agricultural land size, and urban built-up areas, while other factors, such as soil characteristics and topography, were not considered. Including these additional variables could provide a more comprehensive understanding of surface runoff dynamics. Furthermore, the study's applicability may be limited to the specific geographic and climatic conditions of the study area. Extrapolation of the findings to other regions should be done with caution, as hydrologic processes can vary significantly across different regions. Overall, the findings of the runoff sensitivity analysis can be helpful in decision-making processes related to land use planning, water resource conservation, and sustainable land management practices.

7 Upstream–Downstream Hydrologic Connectivity and Delta Implications

7.1 Prelude

Rivers function as dynamic, interconnected systems that span from their headwaters to their mouths, forming what is often termed a "source-to-sink" continuum. This continuum reflects the spatial and functional linkage between upstream and downstream environments. Within this framework, upstream–downstream linkages represent a critical dimension of river basin systems, where physical, ecological, and socio-economic processes are deeply interwoven (Syvitski et al. 2009; Best 2019).

In theory, river basins should be managed holistically to reflect this interconnectedness. However, the extent and magnitude of these linkages are frequently underappreciated in practice. Fragmented governance structures, jurisdictional boundaries, and sectoral policies often lead to disjointed planning and interventions that fail to consider how upstream changes cascade downstream. As a result, activities such as upstream deforestation, mining, or dam construction may inadvertently impair delta morphology, water quality, sediment balance, and ecological resilience in the deltaic regions (Meybeck and Vörösmarty 2005; Kondolf et al. 2014; Poff and Schmidt 2016).

In the South Koel River Basin, earlier chapters have shown how rainfall variability, land use change, and human interventions have transformed the hydrologic regime. These transformations extend beyond the basin itself, affecting the Brahmani–Baitarani delta to which the South Koel contributes water, sediment, and nutrients (Fig. 7.1). Reductions in runoff, sediment deficit, and changes in nitrogen supply have implications for delta health, resilience, and sustainability.

This chapter builds on the analyses of hydro-meteorological variability (Chap. 4), quantified the relative impacts of climatic variability and human activities (Chap. 5), and assessed land use-induced sensitivity in runoff patterns (Chap. 6). It places these findings within a broader upstream–downstream framework, showing how local basin changes

S. Mondal and R. K. Paul, *Climate Variability and Human Interference*, Deltas of the World, https://doi.org/10.1007/978-3-032-13257-4_7

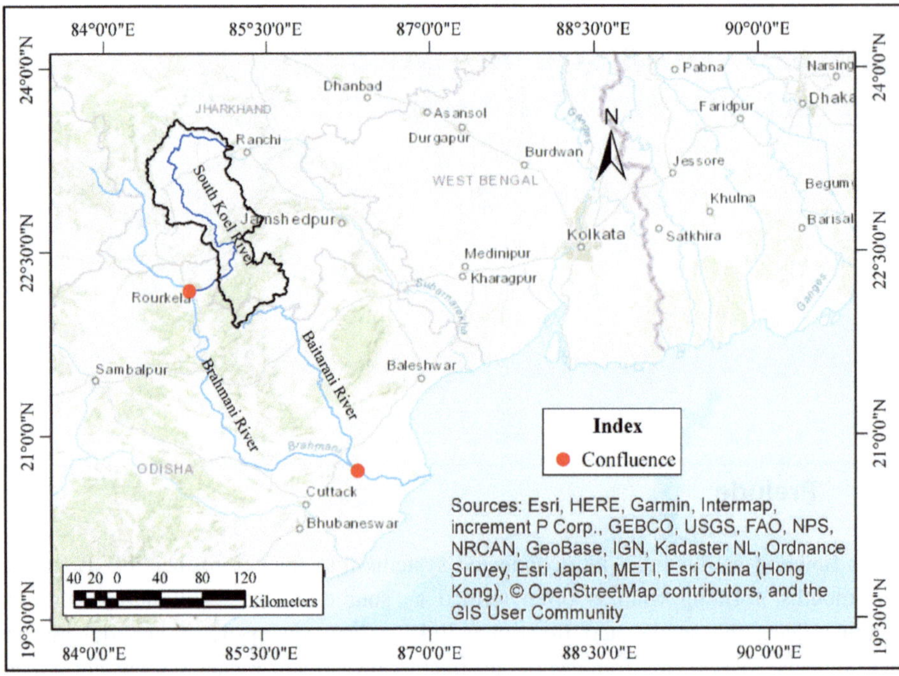

Fig. 7.1 Upstream–downstream hydrologic connectivity and Brahmani–Baitarani delta system

accumulate into large-scale environmental challenges. To analyze these linkages, datasets on runoff, sediment, and nitrogen transport-derived from trend analyses and SWAT model simulations-are employed as the input database to assess hydrologic connectivity between the South Koel Basin and the Brahmani–Baitarani deltaic system.

7.2 Database and Methods

The database used in this chapter builds upon the runoff reduction findings as mentioned earlier, which served as a key input for analyzing upstream–downstream hydrologic connectivity. Using the SWAT model outputs (1981–2018), additional datasets on sediment yield and nitrogen were generated to support the assessment of basin-to-delta linkages. Sediment data and nitrogen-related data were extracted from model simulations calibrated for the study period. These datasets formed the basis for evaluating the influence of upstream hydrologic alterations on downstream deltaic systems.

The runoff reduction results were integrated with sediment and nitrogen data to deduce upstream-driven impacts on the Brahmani–Baitarani delta. For sediment yield analysis, autocorrelation in the time series was first examined, followed by the Modified Mann–Kendall Test, which is appropriate for autocorrelated datasets. This approach allowed the detection of trends in sediment delivery over the study period. For nutrient transport, SWAT-simulated nitrogen data were used to analyze temporal variations and infer potential downstream loading. These analyses provide a comprehensive understanding of how hydrologic changes in the South Koel Basin propagate downstream, affecting sediment continuity and nutrient flux into the deltaic environment.

7.3 Linkages from the South Koel Basin to Brahmani–Baitarani Deltaic System

The South Koel River Basin, though geographically removed from the coastal deltaic plains, maintains an integral hydrological connection to the Brahmani River system, which ultimately contributes to the Brahmani–Baitarani delta and discharges into the Bay of Bengal. This upstream–downstream continuity makes the South Koel Basin a critical determinant in the hydro-ecological functioning of the delta (Fig. 7.2). Changes in upstream hydro-meteorological regimes, significant land use and land cover alterations, and altered runoff sensitivity directly influence sediment delivery, freshwater inflow, and salinity dynamics in the delta.

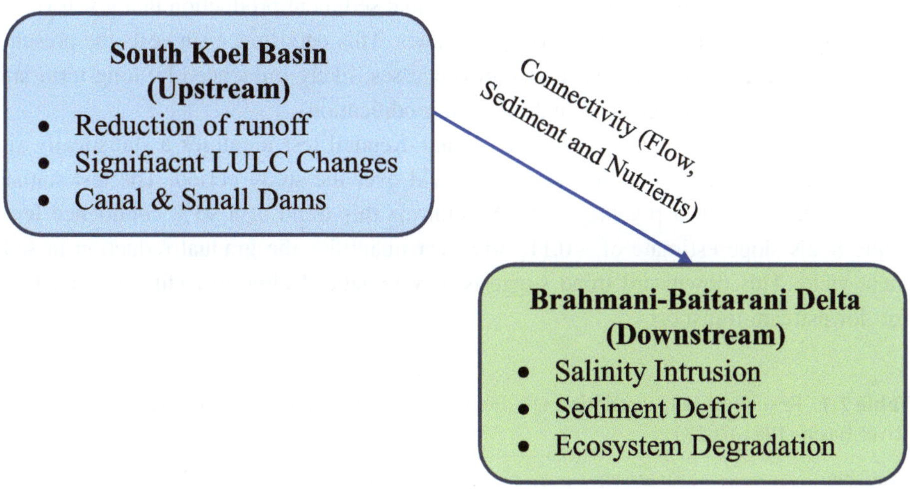

Fig. 7.2 Linkages between the South Koel Basin (upstream) and the Brahmani–Baitarani Delta (downstream)

7.3.1 Runoff Reduction

The temporal changes in runoff characteristics, including the reduction in runoff, have substantial implications for deltaic flow regimes. Altered seasonal flow not only reduces the availability of freshwater downstream but also weakens the dilution capacity of the environment, thereby intensifying salinity intrusion during lean flow months.

Ecological systems of Bhitarkanika mangroves located in Brahmani–Baitarani Deltaic region which depend on a delicate balance between freshwater inflow and tidal influence, become increasingly vulnerable under these conditions. Freshwater scarcity disrupts the salinity gradient, potentially causing species displacement, mangrove dieback, and altered nutrient cycling, as highlighted in global studies on similar tropical deltaic systems (Stanley and Warne 1998; Wang et al. 2007).

7.3.2 Sediment Deficit

The analysis of sediment yield (t/ha) for the South Koel River Basin from 1981 to 2018 reveals significant insights into long-term sediment dynamics under changing land use and land cover and hydrological interventions.

Table 7.1 presents the results of the autocorrelation test using Pearson's product-moment correlation with a lag-1 series. The annual sediment yield exhibits a moderate positive autocorrelation ($r = 0.530$), indicating that sediment production in a given year is positively influenced by its immediate past values. This persistence suggests the presence of a sustained trend in sediment transport processes, likely influenced by long-term land use and land cover changes and hydrological modifications.

As shown in Table 7.2, the Modified Mann–Kendall test identifies a statistically significant negative trend in annual sediment yield over the study period. The test statistic ($Zc = -2.637$) with a p-value of 0.008 confirms this trend at a 95% confidence level, while Sen's slope estimate of -0.113 t/ha/year quantifies the gradual reduction in sediment yield. This downward trend suggests a systematic decline in sediment availability for downstream transport.

Table 7.1 Results of autocorrelation test for year-wise sediment yield (t/ha) for the South Koel River Basin (1981–2018)

Variable	Time-step	r*	*p-value*	Remarks
Sed yield (t/ha)	Annual	0.530	<0.001	Autocorrelated series

*Pearson's product moment correalation with lag 1 series
Source Computed by the authors

Table 7.2 Results of trend analysis for sediment yield (t/ha) for the South Koel River Basin (1981–2018)

Time-step	Modified Mann–Kendall test			Remarks
	Zc	*p*-value	Sen's slope	
1981–2018	−2.637	0.008	−0.113	Significant

Source Computed by the authors

The observed decline in sediment yield from the South Koel Basin aligns closely with prior research on sediment trapping and hydrological alterations in the Brahmani–Baitarani river system. Mishra et al. (2021), using satellite remote sensing and GIS, found that the Rengali Reservoir traps ~50% of the Brahmani River's sediment load, severely diminishing sediment delivery to the deltaic plains. Although the Rengali Dam lies downstream of the South Koel-Sankh confluence, the contribution of sediment from the South Koel plays a crucial role in feeding the Brahmani system. Hence, reduced sediment yield from the South Koel upstream amplifies sediment starvation at the delta.

Choudhury et al. (2007) further confirmed that hydrologic alterations, including upstream damming and catchment changes, have impaired sediment continuity and weakened environmental flows in the lower Baitarani Basin, which is geomorphologically and hydrologically interconnected with the Brahmani–Baitarani delta.

The sediment yield reduction from the South Koel Basin contributes to the broader sediment starvation of the Brahmani–Baitarani delta. In the absence of replenishing sediment loads, the delta becomes increasingly vulnerable to storm surges, sea-level rise, and saline intrusion, especially in low-lying coastal zones such as Bhitarkanika.

7.3.3 Nitrogen Transport

Nitrogen transport from agricultural landscapes in the South Koel River Basin is a critical but understudied process influencing downstream water quality. Using the spatially distributed SWAT simulation model, nitrate (NO_3) movement has four key components: surface runoff (SURQ), lateral flow (LATQ), percolation (PERC), and crop uptake (CROP%).

The simulation revealed that nitrate transport is primarily governed by percolation and surface runoff. Percolation losses exceeded 30 kg/ha, with a maximum value reaching 35.6 kg/ha, indicating substantial nitrate leaching into the vadose zone and shallow aquifers. These deep percolation losses represent a long-term environmental threat, as nitrate that bypasses the root zone may re-emerge through groundwater baseflow, irrigation return flow, or seepage, ultimately reaching river channels and deltaic system.

Surface runoff pathways contributed moderately to nitrate transport, with values reaching up to 2.69 kg/ha, posing short-term but acute risks of nutrient enrichment in surface waters, particularly during heavy rainfall or irrigation events. Lateral subsurface flows (interflow), while quantitatively lower (generally below 0.5 kg/ha), are still important in hilly and well-drained zones, acting as a persistent vector for nitrogen to enter stream networks during low-flow conditions.

Crop nitrogen uptake efficiency across the basin varied from 33% to over 70%, indicating heterogeneity in fertilizer use efficiency. Basin areas with lower uptake percentages (<50%) are more vulnerable to nitrogen losses, either through runoff or deep leaching, suggesting that nutrient application rates exceed agronomic demand in those areas.

The cumulative effect of nitrogen transport from the South Koel Basin contributes to the enrichment of downstream water bodies, including the Brahmani and Baitarani rivers and their joint deltaic system. In particular, the Bhitarkanika mangrove zone, located at the confluence of these rivers and the Bay of Bengal, is highly sensitive to nutrient loading. Excessive nitrate inflow promotes eutrophication, alters species composition, and disrupts natural ecological balances.

The predominance of percolation as a nitrate transport pathway also raises concern for groundwater quality degradation, which may further impact drinking water safety, irrigation viability, and baseflow chemistry during dry seasons.

7.4 Evidence from Global and Regional Case Studies

The upstream–downstream linkages observed in the South Koel River Basin are not unique. Around the world, similar transformations have been documented, particularly in large river systems where human interventions have disrupted the natural continuity of hydrologic and sediment processes.

In the Nile River (Egypt), the construction of the Aswan High Dam has drastically reduced downstream sediment delivery, leading to coastal erosion and declining fertility in the Nile Delta. Despite improved water regulation for irrigation and hydropower, the dam has disrupted the river's natural flood pulse and nutrient flux, illustrating the trade-offs involved in upstream interventions (Stanley and Warne 1998).

The Mississippi River system has undergone extensive channelization and dam construction, resulting in a substantial decline in sediment supply to the delta. As a result, the Mississippi Delta now faces severe land loss and saltwater intrusion (Blum and Roberts 2009). This example highlights how upstream engineering can trigger long-term geomorphic and ecological degradation in deltaic regions.

In Asia, the Ganges–Brahmaputra-Meghna (GBM) system demonstrates complex transboundary upstream–downstream tensions. Dam construction, deforestation, and shifting monsoon dynamics have affected sediment transport and flood patterns, with downstream consequences for agricultural productivity, riverbank erosion, and saline water intrusion in the Bengal Delta (Rahman et al. 2018).

The Mekong River further underscores the challenges of cumulative upstream interventions. With multiple dams constructed or planned across the basin, sediment load reaching the Mekong Delta has significantly decreased. This has led to a rise in relative sea level, bank erosion, and reduced agricultural yields, especially in Vietnam's fertile delta plains (Kondolf et al. 2014b).

In China's Yellow River, aggressive water abstraction, deforestation, and damming have significantly decreased downstream flow and sediment discharge, leading to increased deltaic vulnerability (Wang et al. 2007).

These case studies illustrate a common narrative: upstream interventions-whether for flood control, hydropower, or water diversion-often generate unintended downstream consequences. These include sediment starvation, altered flow regimes, reduced ecological resilience, and deltaic degradation.

7.5 Policy and Management Implications

The findings from this study underscore the critical need for integrated basin-to-delta water governance that recognizes upstream interventions as key drivers of downstream hydrologic and ecological outcomes. The South Koel–Brahmani system exemplifies how localized land use and land cover change, and hydrologic modifications can cascade into broader deltaic vulnerabilities, including saline intrusion, wetland degradation, and aquaculture disruption.

7.5.1 Integrated River Basin Management (IRBM)

There is a pressing need to implement a basin-wide, multistate IRBM framework for the Brahmani–Baitarani basin, aligning with the Draft River Basin Management Bill (2018). Institutional mechanisms should enable data sharing, transboundary coordination, and joint decision-making across Jharkhand and Odisha.

7.5.2 Ecological Flow (E-Flow) Regulation

Maintaining environmental flows is essential to preserve salinity balance and deltaic ecosystem health. Flow thresholds during dry seasons must be legislated and enforced.

7.5.3 Land Use Zoning and Catchment Restoration

Zoning regulations should limit forest-to-agriculture conversion and promote catchment restoration in erosion-prone zones. Payment for Ecosystem Services (PES) schemes may be piloted to incentivize conservation. Furthermore, precision fertilizer application, improved crop uptake efficiency, and vegetative buffer strips to mitigate nitrogen losses. Moreover, monitoring and regulating fertilizer use in nitrate-vulnerable zones of the South Koel Basin is crucial to reduce the nutrient burden on the Brahmani–Baitarani delta.

7.5.4 Sediment Budget Monitoring

Periodic sediment monitoring and modeling should be institutionalized, especially in reservoir zones of the delta. Sediment bypassing or flushing protocols should be explored to support delta building processes.

7.5.5 Climate-Resilient Delta Governance

Management strategies must incorporate climate risk modeling, including sea-level rise, extreme rainfall variability, and estuarine salinity projections. This requires linking hydrologic models (e.g., SWAT) with deltaic and estuarine models (e.g., Delft3D, MIKE21).

7.5.6 Community-Based Co-management

Local stakeholders particularly farmers, fisherfolk, and tribal forest dwellers must be involved in flow planning, wetland management, and climate adaptation initiatives. Grassroots water user associations can play a key role in maintaining distributary channels and embankments.

7.6 Concluding Remarks

This chapter has established a critical upstream–downstream linkage between the South Koel River and the Brahmani–Baitarani deltaic system, emphasizing that hydrologic transformations in upland regions have tangible, often adverse, impacts on deltaic dynamics. Drawing on empirical trends, sediment insights, and modeling frameworks, the chapter demonstrates how reduced runoff, sediment deficit, and altered flow regimes driven by primarily anthropogenic factors compromise the ecological stability and livelihood security of downstream zones.

Evidence from global and regional delta systems reinforces that these processes are neither isolated nor hypothetical. Instead, they are part of a larger pattern of deltaic vulnerability under catchment stress. The South Koel Basin represents a microcosm of this global phenomenon, where unchecked land use and land cover change, flow regulation, and basin mismanagement can lead to cumulative degradation downstream.

To address these emerging challenges, a systems-based governance approach anchored in integrated river basin planning, sediment management, e-flow regulation, and community participation is essential. The findings of this chapter lay the foundation for forward-looking policy interventions that aim to ensure resilience of both headwaters and deltas under accelerating climate and socio-economic transitions.

8 Major Findings and Implications

8.1 Prelude

This chapter brings together the major findings of the study, distilling the results into a coherent narrative that links climate variability, land use change, and hydrologic responses in the South Koel River Basin. While earlier chapters explored rainfall-runoff dynamics, quantified the roles of climate and human activities, examined land cover sensitivity, and established upstream–downstream linkages, this chapter synthesizes those insights to reveal the broader picture of basin transformation.

The discussion moves beyond isolated results to show how the observed decline in runoff, sediment deficit, and nutrient imbalances reflect a deeper shift in basin function under human pressure. Objective-wise findings are consolidated and interpreted, highlighting both empirical evidence and conceptual contributions. By framing these outcomes within regional and global debates, the chapter identifies the knowledge gaps addressed, the methodological advances made, and the implications for sustainable water governance. In doing so, it sets the stage for translating scientific understanding into actionable strategies.

8.2 Summary of the Work Design

This research assessed the impact of climate variability and human activities on the alteration of runoff in the South Koel River basin in Eastern India. There are four specific objectives: (i) to assess the temporal variability of rainfall and runoff since 1973 in terms of trend and step change in South Koel River Basin; (ii) to determine the relative contribution of climate variability and human activities to cause the alteration of runoff in the study area; (iii) to capture the runoff sensitivity to changes in individual land use and land

S. Mondal and R. K. Paul, *Climate Variability and Human Interference*, Deltas of the World, https://doi.org/10.1007/978-3-032-13257-4_8

cover class with special reference to forest cover, agricultural land use, and urban built-up area in the concerned river basin, and (iv) to establish the hydrologic linkages between the upstream South Koel Basin and downstream Brahmani–Baitarani deltaic system.

With these four objectives, an attempt was made to find the answers to four specific questions:

(i) What type of changes have occurred in the stream flow: gradual change (trend) or step change or both?
(ii) How much change in runoff is due to climate variability and how much is due to human activities?
(iii) What is the threshold change value in forest cover, agricultural land, and urban built-up area that has a functional impact on runoff alteration?
(iv) How do upstream hydrologic transformations in the South Koel Basin influence downstream deltaic systems, particularly the Brahmani–Baitarani delta?

As per the *first objective*, rainfall and runoff are the response variable that were analysed with respect to time. Here, daily rainfall data were sourced from NASA power data access, Langley Research Centre (NASA 2020) for the period of 1981–2018 the grid points with a grid resolution of $0.625° \times 0.50°$ whereas daily discharge data were obtained from the Department of Water Resources, River Development, and Ganga Rejuvenation, Ministry of Jal Shakti, Government of India (India-WRIS 2020) for the Jarikela gauge station for the period of 1973–2018. The daily data were transformed into year-wise total seasonal and annual rainfall (mm) and runoff series (mm), which were analysed for step change and trend.

For rainfall, step change and trend were analysed for the period of 1981–2018 since data were available for this period only, whereas for runoff, these tests were done for the period 1973–2018. Again, the runoff was analysed for the period 1981–2018 so that the results of runoff and rainfall could be compared, which is needed to fulfil the second objective. However, for making the seasonal series of rainfall and runoff data, IMD-defined seasons have been followed. In India, IMD identified four meteorological seasons: (a) pre-monsoon (March–May), (b) Monsoon (June–September), (c) post-monsoon (October–November), and (d) winter (December–February).

For step change analysis, the Buishand U, the Buishand Range, and standard normal homogeneity tests were applied for normal datasets, while Pettitt's test and sequential Mann-Kendall test were used. For trend analysis, the Mann–Kendall test was applied to data without autocorrelation, while the Modified Mann–Kendall test was used for autocorrelated datasets. The magnitude of the trend was estimated by using Sen's Slope.

Normality was assessed using Kolmogorov–Smirnov and Shapiro–Wilk tests, lag1 autocorrelation was checked by using Pearson's r, and the significance of r was tested by t-Test. However, the significance level for hypotheses tests was set at 5%.

As per the *second objective*, runoff is the response variable, whereas climate variability and human activities are the explanatory variables. Land use and land cover was taken as the representative of human activity. However, the relative contribution of climate variability and human activities on the runoff alteration was assessed using the SWAT model in the ArcSWAT extension of Arc GIS software.

The SWAT model was set up using the SRTM DEM, soil, and LULC datasets, then run with the weather datasets to simulate the annual runoff. The DEM dataset was collected from the USGS Earth Explorer (USGS Earth Explorer 2022) to calculate flow direction, generate the stream networks, delineate the watershed, and prepare the slope map.

Soil datasets were downloaded from the https://soilgrids.org/ website (ISRIC-WDC 2020) and formatted as per the SWAT model's specifications (textual form). Some of the soil data were readily available, while some of them needed to be prepared. The datasets on the available water capacity, soil hydraulic conductivity, moist soil albedo, and USLE soil erodibility factor were prepared using pedo-transfer functions. The readily available datasets were soil hydrologic group, maximum rooting depth of the soil profile, depth from the soil surface to bottom of the layer, moist bulk density, organic carbon content, clay content, silt content, sand content, and rock fragment content. Soil textural layer in raster form was also required, which was prepared using the USDA guidelines.

Two LULC maps were prepared for the years 1995 and 2016 based on the Landsat TM and Landsat OLI satellite images, respectively, which were collected from the USGS earth explorer website (USGS Earth Explorer 2022).

Daily weather datasets, namely daily rainfall, maximum and minimum temperature, relative humidity, wind speed, and solar radiation data, were collected from the NASA power data access, Langley Research Centre (NASA 2020) for the period of 1981–2018 having grid point resolution of $0.625° \times 0.50°$. The missing values were filled by the SWAT's default value of $-$ 99. These datasets were used in the weather generator (WGN) parameters estimation tool, which prepared the weather datasets as per the SWAT model.

The daily runoff data was collected from the same source as used to solve the first objective. The runoff dataset was used only for the validation and calibration purposes of the SWAT model.

After the preparation of the necessary datasets as per the SWAT model's specifications, these datasets were ready for the process of runoff simulation. The runoff was simulated based on the findings of the first objective. The Buishand U test and Buishand Range test both detected the same change point in 2001, whereas the standard normal homogeneity test detected a change point in 2008 for the annual runoff data series. Based on these findings, the whole study period (1981–2018) was divided into the reference period and the interference period. The change point 2001 divided the study period into the reference period 1981–2001 and the interference period 2002–2018. The change point 2008, on the other hand, divided the study period into the reference period 1981–2008 and the interference period 2009–2018.

However, a total of six simulations (Sim) models were performed, three for each change point, in order to fulfil the second objective. Sim1 was run by LULC and weather datasets of the reference period, Sim2 was run by LULC and weather datasets of the interference period, and Sim3 was run by LULC of the reference period and weather datasets of the interference period. So, any difference between the mean runoff of Sim2 and Sim1 is the total change in runoff caused by the combined effect of climate variability and LULC change. The difference between Sim3 and Sim1 describes the change in runoff caused by climate variability, whereas the difference between Sim2 and Sim3 is due to the influence of LULC change alone.

The SWAT model was calibrated by using SUFI-2 algorithms in SWAT-CUP software. The model validation was also done properly. Two objective functions—NSE and R^2—were used to judge the level of calibration and validation. Subsequently, the relative sensitivity of input parameters was also analysed.

After the assessment of the relative contribution of climate variability and human activities on the alteration of runoff, the significance of the difference in SWAT runoff output between the two simulations was judged by using an independent t-Test and a necessary normality check for SWAT output runoff series was checked by using Kolmogorov–Smirnov and Shapiro–Wilk tests. Cohen's d was used for estimating the effect size for those series that showed significant differences in the t-Test.

As per the *third objective*, there are three sets of variables. In one set, surface runoff is the response variable; forest cover is the fixed factor with four specific size classes of forest cover, and rainfall is the covariate. In another set, again surface runoff is the response variable, agricultural land is the fixed factor with four classes of agricultural land based on areal coverage, and rainfall is again the covariate. In the third set, runoff is the response variable, the urban built-up area is the fixed factor with four specific built-up area classes made based on the size of the built-up area, and rainfall is the covariate.

These datasets were generated by the SWAT simulation model for the period of 1981–2001. Three separate one-way ANCOVA models were run on the above-mentioned three respective datasets to determine whether the mean runoff varies across the four size classes of LULC, namely forest land, agricultural land, and urban built-up area. The aim is to determine the threshold value of the areal unit of each LULC class above and below which the runoff changes significantly. ANCOVA is the best-suited tool here for this kind of analysis in which rainfall is controlled, but along with the ANCOVA model, here, One-way ANOVA was also used to know whether any difference is created in the outcome, i.e., to know the output difference with or without controlling the rainfall effect.

As per the fourth objective, the study explored the hydrologic connectivity between the upstream South Koel Basin and the downstream Brahmani-Baitarani deltaic system, with a focus on runoff variability, sediment delivery, and nitrogen flux. The observed changes in runoff such as reduced surface runoff was analyzed in relation to potential impacts on freshwater inflow and salinity balance. Sediment delivery to the delta was assessed based on the SWAT simulation model. Global and regional case studies were reviewed to contextualize these impacts, highlighting how diminished sediment load and altered flow regimes contribute to ecological degradation, salinization, and delta vulnerability. The analysis emphasized that changes in the upland basin, particularly post-2000, have cascading downstream consequences, reinforcing the need for integrated upstream–downstream hydrologic planning and adaptive delta governance frameworks.

8.2.1 Thematic Interpretation of Findings

The study reveals that hydrological change in the South Koel River Basin is significantly shaped by how land is used, modified, and managed over time. While climatic variability contributes to insignificant changes in runoff patterns, it is the intensification of human interventions mainly forest degradation, and agricultural expansion that have had a more consistent and measurable impact, especially after the year 2000.

The disproportionate influence of land use and land cover change on runoff responses indicates that watershed-scale planning must integrate terrestrial dynamics as co-drivers of hydrological alteration, not treat them as secondary to climate. The SWAT model simulations confirmed that even changes in land use and land cover particularly in forest cover and agricultural land can shift runoff behavior in ways that impact both upstream water availability and downstream deltaic stability.

The findings of the study underscore the importance of preserving ecological buffers such as forested uplands, regulating intensive agriculture, and promoting land zoning frameworks in hydrologically sensitive zones. The results point toward an urgent need for catchment-to-delta integrated management strategies, recognizing that changes in upland processes cascade into downstream estuarine stress, including sediment deficit, salinity intrusion, and wetland degradation.

These thematic insights directly align with the Sustainable Development Goals (SDGs) notably:

- SDG 6 (Clean Water and Sanitation), through improved watershed governance;
- SDG 13 (Climate Action), by informing adaptive hydrologic planning; and
- SDG 15 (Life on Land), by emphasizing land restoration and ecosystem services.

8.3 Objective-Wise Major Findings

The obejective wise findings are presented in Table 8.1.

Table 8.1 Summary of findings at a glance

Objective	Key methods used	Main findings	Implication
Objective 1 (Chapter 4)	Mann–Kendall, Pettitt Test	Rainfall has no significant trend; runoff decreased significantly post-2001	Indicates strong anthropogenic stress
Objective 2 (Chapter 5)	SWAT model simulations, Cohen's d	LULC responsible for ~ 73% runoff reduction; climate for ~ 27%	Human activity is the dominant factor
Objective 3 (Chapter 6)	ANCOVA	Runoff is sensitive to forest (> 1.35 km^2) and agriculture (> 1.30 km^2); urban area has minimal effect	Identifies land cover thresholds for planning
Objective 4 (Chapter 7)	Statistical techniques, SWAT Model and literature synthesis	Upstream runoff reduction, sediment deficit linked to deltaic salinity intrusion and erosion risks, nitrogen supply leads to degraded deltaic health	Highlights need for integrated upstream–downstream water and sediment governance frameworks

Source: Computed and deduced by authors

(a) ***First Objective:*** *To assess the temporal variability of rainfall and runoff since 1973 in terms of trend and step change in South Koel River Basin.*

(i) *Findings on Step Change of Rainfall and Runoff*

No significant changes were found in any of the seasonal and annual rainfall series, either the basin average series or the series specific to the grid point for the period 1981–2018. Contrary to this, significant change points were detected for the monsoon and annual runoff series for the period 1981–2018. The Buishand U and Buishand Range test detected 2001 as the change point, whereas the Standard Normal Homogeneity test identified 2008 as the change point for both the series, i.e., monsoon and annual one. The same happens to the monsoon and annual series for the period 1973–2018.

(ii) *Findings on Trends in Rainfall and Runoff*

No significant trend was found for any of the basin average seasonal and annual rainfall series for the period 1981–2018. The same happens to the grid-specific rainfall series with an exception in grid points 14, 19, and 23. The annual series shows a significant increasing trend at grid point 14 (at the rate of 8.46 mm/year) and 19 (at the rate of 9.523 mm/year), whereas the pre-monsoon series at grid point 14 shows an increasing trend (at the rate of 1.29 mm/year). The grid point 23, on the other hand, shows an increasing trend for the monsoon series (at the rate of 5.004 mm/year). It was observed that the oscillatory behavior in the rainfall time series, in most cases, makes the whole period series trendless. During the pre-monsoon, the basin average rainfall shows an increasing tendency since 1991 onward, while the winter rainfall was found to have decreasing tendency up to 1993 and trendless thereafter.

Runoff, on the other hand, shows a decreasing trend for the monsoon and annual series for the period 1981–2018. The rate of decrease for the monsoon series is 5.29 mm/year, while that for the annual series is 5.929 mm/year. The same tendency is observed for the monsoon and annual series for the period 1973–2018, although the magnitude of the trend is somehow different. Monsoon runoff during 1973–2018 shows a decreasing trend at the rate of 4.731 mm/year, while the annual series during the same period had decreased at the rate of 5.126 mm/year. There are sufficient substantial evidences (Table 4.17) to say that the oscillatory temporal behavior of pre-monsoon, post-monsoon, and winter runoff series on decadal-scale during have made them trendless for the whole study period.

Note on Achieving First Objective

Whatever findings were obtained from step change and trend analysis of rainfall and runoff series are found to be very much consistent with each other.

The interesting feature which could be deduced from the current analysis is that the monsoon and annual runoff series are showing a significantly decreasing trend when the rainfall series for the corresponding time step is found to be trendless. So, the decreasing runoff tendency could be attributed to human interference only. This finding would be substantiated by the analysis of the impact of human interference on runoff alteration, which is incorporated in the second research objective that is discussed in the next subsection. Through field investigation, it was clear that human intervention in the form of water diversion and changes in agricultural technology are very much responsible for this kind of trend in runoff when rainfall is found to be trendless. The findings in connection to the second objective, which is discussed in the next subsection, will provide ample evidence regarding human intervention that could be used to substantiate the present findings deduced in connection to the first objective.

So, the clarity of the findings is satisfactory enough to say that the first objective of this research has been fully achieved.

(b) ***Second Objective:*** *To assess the relative contribution of climate variability and human activities to cause the alteration of runoff in the study area.*

(i) *Findings on Relative Contribution of Climate Variability and LULC Change*

Based on 2001 as the change point, the whole study period was divided into the reference period between 1981–2001 and the interference period between 2002–2018. The mean annual runoff decreased by 195.87 mm from reference to the interference period due to the combined effect of climate variability and change in LULC. But the individual role of both factors is found to be just opposite to each other. The LULC caused a decrease in runoff in between the above-mentioned period by 280.15 mm, whereas climate was found to increase the runoff by 84.29 mm. In terms of percentage to total change, LULC caused a decrease in runoff by 143.03%, and the climate, on the other hand, caused an increase in runoff by 43.03%.

Based on 2008 as the change point, the defined reference period is 1981–2008, and the interference period is 2009–2018. In this context, the same response was observed for the runoff series, although the magnitude of change is numerically different. The total decrease in runoff was 224.64 mm from reference to the interference period due to the combined effect of climate variability and LULC change. The LULC caused a decrease in the runoff by 282.77 mm, whereas climate variability caused an increase in the runoff by 58.13 mm between the reference and interference periods. The percentage share of LULC to decrease in runoff is 125.88%, and that of climate variability to increase the runoff is 25.88%.

(ii) *Judging the Significance of Relative Contribution*

For the reference and interference period as defined by change point 2001, the relative contribution of climate variability was found to be insignificant, whereas that of LULC was significant at a 5% significance level. Their combined effect was also significant at a 5% significance level. The same findings were observed for the reference and interference period defined based on 2008 as the change point, but the effect size of the change in runoff for both the 2001 and 2008 defined reference and interference period were not numerically the same. In the context of 2001 as the change point, the effect size of LULC was 2.27, which is treated as huge effect size (Cohen 1988), whereas in the context of

2008 as the change point, the effect size of LULC contribution was 1.95 which is treated as very large effect size (Cohen 1988).

Mentioning the effect size of climate variability is irrelevant since it was found to be insignificant. But the effect size of their combined influence as per 2001 defined reference and interference period was 1.43, whereas based on the 2008 defined period, it was 1.50, which is also a very large one (Cohen 1988).

Note on Achieving Second Objective
The relative contribution of climate variability and human activities on the alteration of runoff was properly separated out. The findings that were deduced in the context of the first objective were that rainfall is found to be trendless, whereas runoff is showing a significant decreasing trend which suggests that human intervention is the main driving force behind such change. The current findings in the context of the second objective substantiate the previous ones. Another important thing to be mentioned here is that LULC change analysis revealed that there was a decrease in forest cover by about 7.04% between 1995 and 2016, whereas during the same period, there was an increase in agricultural land by about 4.25%, which together are supposed to increase the runoff, but in first and second objective it is observed that instead of increase, the runoff is decreasing. This could be attributed to changes in agricultural technology. Through field investigation, it was known that in the last 15–20 years there had been introduced more water-consuming high-yielding variety (HYV) seeds which are playing an effective role in reducing runoff.

So, the evidences are strong enough to substantiate the findings that have been obtained in connection to the second objective. So, the relative contribution of climate variability and LULC change on runoff alteration could be accepted as valid and reliable findings. In this regard, it can be said that the second objective of the present research has been fulfilled successfully.

(c) ***Third Objective:*** *To capture the runoff sensitivity to changes in individual land use and land cover classes with special reference to forest cover, agricultural land use, and urban built-up area in the concerned river basin.*

(i) *Findings on Runoff Sensitivity to Forest Cover*

The ANOVA model, which analysed the variance in runoff without controlling rainfall, shows that mean runoff changes significantly across the size classes of forested land, but it is not equally responsive to each size class of forest cover. The significant change in runoff occurs if the size of forested HRU exceeds 1.35 km^2 and significant changes are observed in between the forest cover above and below the areal coverage of 9.14 km^2. Here, the effect size as measured by partial eta squared is 0.019, i.e., 1.9% of the total variance in runoff is explained by size variation in forested HUR.

The same findings are observed in the ANCOVA model, which analysed the runoff variance after eliminating the effect of rainfall. The partial eta squared value, here, remains the same, which indicates that the rainfall variation within forest land is less important than the variation in areal coverage of forested land in alteration of runoff.

However, the significant runoff change occurs when forest cover goes beyond the value of 1.35 km^2, beyond 3.44, and beyond 9.14 km^2, but there is no significant change in runoff above and below the values of 3.44 and 9.14. It implies that the runoff of the first land class is different from that of the second, third, and fourth one, but in between the land class of the second and third, and between the third and fourth, there is no significant change in the runoff. This finding reveals the value of 1.35 km^2 as the threshold areal coverage value of forest beyond which runoff changes.

(ii) *Findings on Runoff Sensitivity to Agricultural Land*

The ANOVA model for agricultural land produces similar findings as observed for forest cover, but the variance accounted for is found to be different. The variation in the size of agricultural land explains 3.1% of the total variance in the runoff. The ANCOVA model, which analysed the variance after eliminating the rainfall effect, reveals that variation in agricultural land size explained 2.3% of the total variance in the runoff series. The ANOVA model shows that runoff changes significantly if agricultural land size exceeds 1.30 km^2. Again, it differs between the land size above and below 10.5 km^2. The ANCOVA model, on the other hand, shows that runoff differs significantly only when the land size exceeds the value of 1.30 km^2. Hence, the threshold land size value for agricultural land is determined to be 1.30 km^2.

(iii) *Findings on Runoff Sensitivity to Built-Up Area*

As per ANOVA and ANCOVA models output, the runoff is found to be insensitive for the built-up area, which is quite unrealistic. But this might be attributed to the fact that the built-up area occupied only 4.12% in 2016, whereas it was only 0.50% in 1995, and the built-up area is found to have a scattered distribution. Hence, its role is not proven to be important in runoff alteration.

Note on Achieving Third Objective
The relative sensitivity of runoff to changes in forest cover, agricultural land, and urban built-up area was properly assessed and justified. The findings are clear and satisfactory, indicating the successful accomplishment of the third research objective.

(d) ***Fourth Objective:*** *To analyze how upstream hydrologic transformations in the South Koel Basin influence downstream deltaic systems, with emphasis on the Brahmani–Baitarani delta.*

(i) *Findings on Upstream–Downstream Hydrologic Linkages*

The study established clear hydrologic connectivity between the South Koel Basin and the Brahmani-Baitarani delta through multiple lines of evidence derived from flow variability, sediment dynamics, and nutrient flux perspectives. The analysis revealed that the reduction in runoff has significantly impacted downstream freshwater availability in the coastal zones. This altered flow regime reduces the dilution capacity of the region, intensifying salinity intrusion. This phenomenon mirrors observed conditions in the Brahmani River, where upstream discharge reduction has led to ecological degradation, including stress on mangrove ecosystems and a decline in agricultural productivity.

Sediment dynamics further demonstrated the impact of anthropogenic modifications have led to a sediment deficit in the deltaic zone, weakening natural land-building processes. The Modified Mann–Kendall test confirmed a statistically significant negative trend ($Zc = -2.637$, $p = 0.008$), with Sen's slope estimating a reduction of − 0.113 t/ha/year during 1981–2018. Case study comparisons such as Mekong and Mississippi deltas support the argument that reduced sediment supply contributes to delta subsidence, wetland degradation, and coastal erosion.

Nitrogen transport analysis showed that nitrate percolation losses exceeded 30 kg/ha, peaking at 35.6 kg/ha, while surface runoff transport reached up to 2.69 kg/ha, indicating a high risk of nutrient loading during peak flows. Lateral flow contributions remained relatively low (< 0.5 kg/ha) but were relevant in hilly zones for low-flow nitrogen delivery. Crop nitrogen uptake efficiency varied considerably, ranging from 33% to over 70%, with inefficient uptake (< 50%) associated with greater leaching and runoff losses.

Through this objective, the study underscored the importance of viewing river basin systems as longitudinally integrated entities, where upstream changes generate downstream vulnerabilities.

(ii) *Policy and Management Implications*

The findings clearly suggest that the consequences of upstream hydrologic alterations cannot be fully understood or mitigated without incorporating their downstream effects into planning and governance frameworks. The study recommends promoting ecological flow regimes, sediment management strategies, and catchment-to-delta integrated planning, aligning with basin-scale resilience goals. Moreover, institutional coordination across upland and coastal jurisdictions is vital to address the cumulative impacts of altered hydrologic and sediment regimes.

Note on Achieving Fourth Objective
The hydrologic and ecological linkages between the South Koel Basin and the Brahmani-Baitarani delta were successfully established through qualitative, analytical, and comparative evidence. The integration of hydrological findings with deltaic processes provides a compelling case for comprehensive water management that spans the full river continuum. These findings affirm the successful achievement of the fourth research objective and provide a strong foundation for actionable policy frameworks focused on sustainable basin-delta governance.

8.4 Knowledge Contribution and Research Gap Addressed

This study contributes a robust and multi-dimensional perspective to the understanding of basin-scale hydrologic transformation under the dual stressors of climate variability and human interference in a monsoon-driven region of Eastern India. The work fills a critical research gap by providing empirical evidence of hydrologic alteration in the South Koel River Basin, a relatively under-researched catchment with strategic upstream–downstream linkages to the Brahmani-Baitarani delta system. It quantifies the relative contributions of LULC change and climate variability to runoff reduction, introduces threshold-based assessments for forest and agricultural cover, and conceptually establishes sediment and flow connectivity from upland watersheds to downstream deltaic zones.

This integrative approach contributes to the limited but growing literature on compound hydrological stress in monsoon-fed tropical basins and provides a replicable framework for future attribution studies in similar hydro-climatic settings.

8.5 Implications for Science, Policy, and Practice

The findings of this research hold significant value across three intersecting domains: scientific knowledge, policy formulation, and community-level practice.

8.5.1 Scientific Implications

The study advances hydrological science by demonstrating a rigorous methodology combining trend and step change analysis, spatially distributed runoff modeling, and land use and land cover sensitivity assessments within a single basin framework.

The results indicated that land use and land cover (LULC) changes are the dominant factor driving runoff reduction, while the contribution of climate variability is statistically insignificant. This finding challenges the prevailing assumption that climate change is invariably the primary driver of hydrological change in the Global South and establishes a new empirical benchmark for hydrologic attribution studies in tropical basins.

The introduction of threshold values for forest (~ 1.35 km^2) and agricultural land (~ 1.30 km^2) per HRU adds actionable insight into landscape sensitivity. These thresholds offer measurable criteria for land-use regulation and ecosystem service preservation.

Importantly, the study bridges upland hydrological analysis with deltaic response, establishing linkages between runoff decline, sediment deficit, nutrient flux and deltaic vulnerability. This upland–lowland integration is often overlooked in traditional watershed studies and represents a novel thematic contribution.

8.5.2 Policy Implications

The study offers direct implications for river basin governance, land-use planning, and national water resource policy in India.

(a) *Forest Conservation*: Identifying forest cover thresholds with hydrologic significance validates the need to prioritize afforestation and conservation efforts in headwater zones not just for biodiversity but for maintaining downstream flow regimes.
(b) *Agricultural Zoning and Cropping Strategy*: The study reveals that intensive cultivation beyond a certain spatial threshold reduces runoff. This supports policies promoting water-efficient crops, agroecological zoning, and climate-resilient farming.
(c) *Urban Planning*: Although urban areas currently contribute minimally to runoff alteration due to scattered distribution, the findings stress the importance of proactive planning in view of anticipated urban densification, ensuring impermeable surface expansion is controlled.
(d) *Watershed and Deltaic Integration*: The work supports a shift from isolated watershed management to a catchment-to-coast governance model. It aligns with India's Jal Shakti Abhiyan and National Water Policy, and provides scientific backing for integrated management plans linking upland flow regulation with downstream delta resilience.

8.5.3 Practical and Community-Level Implications

At the grassroots level, the study provides operational guidance for local governance bodies, NGOs, watershed committees, and self-help groups working in the basin.

(a) *Community Forest Management (CFM)*: Can adopt the identified forest thresholds as targets for local afforestation schemes and participatory micro-watershed planning.

(b) *Rainwater Harvesting and Recharge*: *Initiatives* should focus on runoff-depressed HRUs, particularly where forest loss or intensive farming has occurred.
(c) *Deltaic Livelihood Support*: In downstream areas, especially in the Brahmani deltaic zone, the study highlights the need for integrating freshwater regulation, saline intrusion monitoring, and mangrove conservation.
(d) *Participatory Monitoring and Education*: The modeling outputs can be repurposed for environmental awareness programs at schools, farmer cooperatives, and youth forums to build climate-resilient decision-making capacity.

The study advances the understanding that hydrologic resilience in a monsoonal basin is shaped by land, water, and people necessitating a multi-scalar, interdisciplinary, and inclusive approach to planning and intervention. It sets a precedent for future research and action on upstream–downstream hydrologic integration under climate stress and anthropogenic change.

8.6 Concluding Remarks

This chapter consolidated the objective-wise findings of the research, offering a cohesive narrative that links the observed hydrological transformations in the South Koel River Basin particularly to anthropogenic stress. The evidence from trend analysis, hydrological modeling, land use sensitivity assessments, and upstream–downstream linkage evaluation strongly confirms that land use and land cover changes particularly deforestation, and agricultural intensification are the predominant drivers of runoff alteration in the basin since the 2000.

While climate variability has introduced temporal fluctuations in rainfall regime, its standalone impact on long-term runoff trends is statistically insignifiant. Instead, the compound effects of land transformation have contributed most significantly to reduced runoff. These human-induced changes have not only altered upstream hydrological regimes but have also begun to manifest as downstream impacts, such as salinity intrusion, sediment deficit, and deltaic health instability in the Brahmani-Baitarani delta system.

The research highlights the context-specific and nonlinear nature of hydrological responses in monsoon-dominated tropical basins. By introducing quantitative thresholds for land use and land cover classes, simulating scenario-based runoff alterations, and articulating upstream–downstream hydrologic connectivity, this study provides a framework for integrated river basin management that bridges science, policy, and practice. The findings underscore the urgency for catchment-based, multiscalar planning approaches, where forest conservation, water-smart agriculture, and participatory watershed governance are jointly addressed.

Ultimately, this research contributes to the evolving discourse on climate resilience and hydrologic sustainability in Eastern India and serves as a reference model for regions facing similar ecological transitions. It opens pathways for evidence-based policymaking, adaptive water governance, and community-informed landscape restoration, setting a precedent for future interdisciplinary research in South Asia's monsoonal river basins.

9 Recommendations and Future Research Directions

9.1 Prelude

The final extract that emerged from the present study is that runoff is decreasing day by day, but the rainfall remains stationary over time. So, it is human interference that effectively causes such alteration in runoff. The agricultural land in the basin has increased at the cost of forest land, and it is supposed to increase runoff, but instead of increasing, the runoff is decreasing. It is due to the use of modern agricultural technology, which requires more water. The high-yielding crop varieties are more water-absorbing ones, and hence, their cultivation causes alteration in the hydrologic system that ensues the reduction in stream flow.

This chapter outlines the policy recommendations in view to restoring the stability in hydrologic system so that sustained flow can be maintained. It also outlines potential directions for future research to strengthen scientific understanding and guide effective water resource management.

9.2 Recommendation

In order to have sustained streamflow, it is recommended to use indigenous agricultural technology in which locally adopted crop varieties are cultivated following the crop rotation, which is well aligned with the seasonality of the local climate. Specific recommendations in this regard are as follows:

(a) Promoting indigenous crop varieties that are naturally adapted to the region's seasonal climate and require less irrigation.

S. Mondal and R. K. Paul, *Climate Variability and Human Interference*, Deltas of the World, https://doi.org/10.1007/978-3-032-13257-4_9

(b) Encouraging crop rotation and mixed cropping systems, which not only conserve water but also improve soil health.
(c) Supporting rainfed agriculture in appropriate areas and strengthening community-based water harvesting systems.
(d) Integrating watershed-based land use planning with water budgeting at the micro-catchment level.

Such measures will not only align agricultural water demand with available supply but also help restore the natural hydrological regime of the basin. These recommendations also complement India's broader sustainability and resilience goals under programs like the Jal Shakti Abhiyan and the National Mission for Sustainable Agriculture (NMSA).

9.3 Further Scope of the Research

The present study has solved the research problem in a scientific manner, and at the same time, it has brought to the forefront many research problems that the current research does not have the scope to resolve.

(a) Since the last four or five decades, there have been significant advancements in the methods used to study the hydrological response of river basins. However, the basic field of inquiry has remained the same, which is the functional relationship between hydrological variables and their causal factors. But the nature of the hydrologic response of basins varies across the space depending upon various factors such as changing climatic conditions, topography, soil characteristics, human activities, and so on. Therefore, the findings of one basin may not be applicable to other basins and thus basin specific investigation is essential to explore the unique nature of the response in order to make proper-planning and rational decisions.
(b) The relationship between human activities and climate change is a complex and multi-faceted phenomenon. Which one is response and which one is explanatory, it depends upon the spatial extent of the area under investigation. This issue has to be fixed before going for investigating the functional relationship between catchment response, and its causal factors, otherwise, the accurate distinction between climate change impact and that of human activities on runoff alteration would be more challenging.
(c) Alterations in runoff can have feedback effects on the environment, such as changes in vegetation cover, soil moisture, and stream habitats, which in turn can further alter the hydrological cycle. Researchers need to account for these feedback mechanisms in their models and develop methods for predicting their effects.

(d) There is a need to use high-quality and long-term databases on hydro-climatic parameters, land use and land cover datasets, soil and rock parameters, etc. in order to fix the issue of uncertainty in simulating the complex processes of runoff responses. Therefore, it is necessary to monitor the hydro-meteorological parameters on a continuous basis and to generate, archive, and disseminate the databases which are not possible without the involvement of government agencies.
(e) The present study assessed the temporal pattern of year-wise seasonal and annual rainfall total and its impact on runoff, but the inter-seasonal and intra-seasonal variability of rainfall and resultant runoff response needs to be incorporated into the investigation.
(f) Extreme events like heavy rainfall within a very short time span and prolonged dry spells after the onset of monsoon, etc., can have significant impacts on streamflow as well as on water resources, but the assessment of all these aspects is beyond the scope of the present study. So, these aspects create the scope of further research.
(g) While this study established upstream–downstream hydrologic linkages between the South Koel Basin and the Brahmani–Baitarani delta, there remains significant scope for quantitative modeling of sediment transport, estuarine salinity dynamics, and deltaic geomorphology. Integrating basin-scale hydrological models (like SWAT) with estuary or delta-specific models (e.g., Delft3D, MIKE21) would enable a more robust understanding of how upstream alterations in flow and sediment load influence downstream coastal systems under varying climatic and anthropogenic pressures.

9.4 Conclusion

The study successfully assessed the impact of climate variability and human activities on runoff alteration in the South Koel River Basin. After achieving all four research objectives, it became evident that runoff is decreasing over time, despite rainfall remaining statistically stationary. This divergence highlights that climate variability is not the principal driver of hydrological change in the basin; instead, land use and land cover (LULC) change particularly the expansion of agriculture at the expense of forest land, coupled with the introduction of more water-intensive agricultural technologies has played a dominant role in altering runoff dynamics.

These results underscore the importance of sustainable land management and context-sensitive water conservation strategies. The sensitivity analysis carried out in the study revealed the threshold effects of forest and agricultural land cover on runoff generation, providing actionable insights for both land use planners and watershed managers. Maintaining optimal forest cover and regulating agricultural intensification through suitable land zoning and crop selection can significantly influence runoff and ensure better hydrological resilience.

Importantly, the study also established functional linkages between upstream runoff alteration in the South Koel Basin and downstream impacts in the Brahmani–Baitarani deltaic system. These upstream changes may contribute to reduced freshwater availability, increased salinity intrusion, and sediment deficit in the delta, potentially jeopardizing both agricultural productivity and deltaic ecosystems. Although this linkage opens up important avenues for future coupled basin-delta system modeling.

The findings are basin-specific, and while they reflect broader processes of climate and human-driven hydrologic transformation, they should not be directly generalized to other basins without context-specific validation. Nonetheless, the present research contributes significantly to understanding runoff alteration in tropical monsoon basins under compound stressors, addressing both national research gaps and contributing to global hydrologic discourse.

Future research should incorporate additional landscape variables such as topography, soil moisture, and vegetation feedbacks as well as extreme event modeling and delta-scale simulations, to further improve understanding and guide evidence-based policymaking for sustainable water resource management in the context of Eastern India and beyond.

Appendix

See Tables A.1 and A.2.

S. Mondal and R. K. Paul, *Climate Variability and Human Interference*, Deltas of the
World, https://doi.org/10.1007/978-3-032-13257-4

Table A.1 Inventory of literature which shows decreasing tendency of runoff

Sl. No.	Country and study period	Study area	Type of watershed	Spatial scale (Km^2)	Climatic region	Elevation (m)	Method	Results (%) (Runoff decreased)		References
								CV	HA	
1	Denmark 1963–1982	Susa catchment and its catchments	Agricultural watershed	1000	Temperate	0–90	T: mk test C.P: N.M M: DPBM & RM	8 l/s (litre/second)	65 l/s (DPBM) 41 l/s (DPBM) & 35 hm^3/year (RM)	(Refsgaard et al. 1989)
2	Ethiopia 1985–1995	Ketar watershed	Agricultural watershed	3220	Tropical region	1700–4000	T: N.M C.P: N.M M: PRMS	Decreased	Decreased	(Legesse et al. 2003)
3	Spain 1945–1995	Central Spanish Pyrenees	Forest watershed	7430	Mediterranean climate	500–3000	T: SRCS C.P: N.M M: LR	70	30	(Beguería et al. 2003)
4	China 1960–1999	Suomo basin of the upper Yangtze	Agricultural watershed	3016	Semi-tropical climate	292–6554	T: N.M C.P: N.M M: TOPMODEL, SWAT & CHARM	60–80	20	(Chen et al. 2005)
5	India 1901–2000	Krishna river basin	Agricultural watershed	258,948	Sub-tropical climate	21–1909	T: N.M C.P: N.M M: STREAM WB model	6–15 decreased	61 decreased	(Bouwer et al. 2006)
6	Central Asia (Kazakhstan and China) 1911–1986	Lake Balkhash basin	Agricultural watershed	413,000	Arid and semi-arid region	342 (mean)	T: N.M C.P: N.M M: NDIC	62	38	(Kezer and Matsuyama 2006)

(continued)

Table A.1 (continued)

Sl. No.	Country and study period	Study area	Type of watershed	Spatial scale (Km^2)	Climatic region	Elevation (m)	Method	Results (%) (Runoff decreased)		References
								CV	HA	
7	China 1961–1997	Wuding River basin, Loess Plateau	Agricultural watershed	30,261	Semi-arid region	600–1800	T: mks test C.P: mks test M: HS model	13	87	(Li et al. 2007a, b)
8	Australia 1960–2000	Snowy River catchment	Forest watershed	7135	Semi-arid region	1–2230	T: N.M C.P: N.M M: SMAR & Ecohydrological models	Decreased	Decreased	(Tuteja et al. 2007)
9	China 1953–2002	Xinjiang River basin, Poyang Lake	Agricultural watershed	15,535	Wet climate zone	50–2200	T: N.M C.P: N.M M: SWAT	More	Less	(Guo et al. 2008)
10	China 1950–2003	Shiyang river basin	Agricultural watershed	41,600	Arid region	1300–5000	T: mk test C.P: Pettitt test M: DCC & MR	Greater in upstream	Greater in downstream	(Huo et al. 2008)
11	USA 1950–1999	Columbia River Colorado River and California River basins	Agricultural watershed	436,620–670,000	Arid region	0–4635	T: N.M C.P: N.M M: VIC	Up to 40 decreased	Up to 60 decreased	(Barnett et al. 2008)

(continued)

Table A.1 (continued)

Sl. No.	Country and study period	Study area	Type of watershed	Spatial scale (Km^2)	Climatic region	Elevation (m)	Method	Results (%) (Runoff decreased)		References
								CV	HA	
12	China 1950–2005	8 catchments of Shiyang River basin	Mixed Watershed	41,600	Arid region	2705–3482	T: Kendall test C.P: Pettitt test M: HS model	8 l/s (litre/second)	65 l/s (DPBM) 41 l/s (DPBM) & 35 hm^3/year (RM)	(Ma et al. 2008)
13	China 1956–2000	11 catchments of Loess Plateau	Agricultural watershed	113,000	Arid and semiarid region	1200–1600	T: mk test C.P: Pettitt test M: HS model	21–57	43–79	(Zhang et al. 2008)
14	USA 1925–2003	Four Large agricultural watersheds of the Midwest USA	Agricultural watershed	3240–8912	Humid continental climate	150–460	T: N.M C.P: N.M M: SRRC Model	Decreased in one watershed	Decreased in one watershed	(Tomer and Schilling 2009)
15	China 1956–2000	9 watersheds of the upper catchment of the Yellow River Basin	Grassland Watershed	222,551	Semi-arid and semi-humid region	1400–6300	T: N.M C.P: Pettitt test M: HS model & DWBS model	Jul-58	42–93	(Zhao et al. 2009)
16	China 1960–1999	8 catchments of Haihe River basin	Agricultural watershed	84,906	Asian Monsoon climate	100–3059	T: mk test C.P: N.M M: RM	Less	More	(Yang and Tian 2009)

(continued)

Table A.1 (continued)

Sl. No.	Country and study period	Study area	Type of watershed	Spatial scale (Km^2)	Climatic region	Elevation (m)	Method	Results (%) (Runoff decreased)		References
								CV	HA	
17	China 1960–2000	Yellow River Basin, China	Agricultural watershed	121,972	Semi humid region	3480–4680	T: N.M C.P: N.M M: Budyko-type (CE model)	30	70	(Zheng et al. 2009)
18	China 1972–2000	Heihe river basin of Loess Plateau	Agricultural watershed	1506	Semi-arid region	522–608	T: mk test C.P: N.M M: SWAT	95.8	9.6	(Li et al. 2009)
19	China 1961–2001	Chao River and Bai River Catchments, Chaobai River of Hai River basin	Agricultural watershed	13,846	Temperate continental monsoon climate	50–2262	T: OC analysis C.P: OC analysis M: DTVGM	Chao-31.4 Bai-29.6	Chao-68.6 Bai-70.4	(Wang et al. 2009a, b)
20	China 1958–2000	Sanchuanhe River basin, Loess plateau	Agricultural watershed	4161	Arid region	650–1900	T: N.M C.P: N.M M: SIMHYD	29.9	70.1	(Wang et al. 2009a, b)
21	China 1961–2000	Yiluo River basin	Grassland Watershed	18,900	Monsoon climate	28,500	T: mk test, YM & LFM C.P: mkt set & YM M: Budyko-type & SWBM	20.07–23.42	76.58–79.93	(Liu et al. 2009)

(continued)

Table A.1 (continued)

Sl. No.	Country and study period	Study area	Type of watershed	Spatial scale (Km^2)	Climatic region	Elevation (m)	Method	Results (%) (Runoff decreased)		References
								CV	HA	
22	Spain 1963–2002	Duero River basin	Agricultural watershed	97,290	Mediterranean climate	700–2500	T: SRCT C.P: N.M M: LR	Decreased	Decreased	(Morán-Tejeda et al. 2010)
23	China 1961–2000	The upper and middle reaches of the Taoer River	Forest watershed	41,600	Semi-arid and semi-humid climate	150–1709	T: mk test C.P: moving t-test M: HA model	45	55	(Li et al. 2010)
24	Australia, South Africa and New Zealand 1952–2005	Australia: 5 catchments South Africa: 1 catchment; New Zealand: 1 catchment	Agricultural watershed	0.04–3.5	Australia: Mediterranean climate (4) and temperate climate (1) South Africa: summer rainfall area New Zealand: mean annual rainfall 1281 mm	Australia: 160–1144 South Africa: 1820–2463 New Zealand: 460–670	T: N.M C.P: Pettit test M: paired catchment, TTA & HS model	16–98 decreased		(Zhao et al. 2010)
25	China 1956–2000	Dongjiang River basin	Forest watershed	35,340	Subtropical humid climate	180–1691	T: mk test C.P: PMW M: BP-ANN	50	50	(Liu et al. 2010)

(continued)

Table A.1 (continued)

Sl. No.	Country and study period	Study area	Type of watershed	Spatial scale (Km^2)	Climatic region	Elevation (m)	Method	Results (%) (Runoff decreased)		References
								CV	HA	
26	China 1950–2000	Baimasi basin of Yellow River	Agricultural watershed	13,915	Semi-arid climate	122–2465	T: N.M C.P: N.M M: VIC	56–89	Nov-44	(Wang et al. 2010)
27	Canada 1953–2006	Forested Willow River watershed	Forest watershed	2860	Summer short and ward and winter long and severe (–40° C)	308–1500	T: N.M C.P: N.M M: MDMC	72.4 mm/ year	58.7 mm/ year	(Wei and Zhang 2010)
28	China 1956–2005	Upstream of Miyun Catchment, Chaobai River of Hai River basin	Agricultural watershed	15,800	Temperate monsoon and semi-arid	50–2292	T: N.M C.P: N.M M: GBHM & CE model	55	41	(Ma et al. 2010)
29	China 1966–1999	Mian Catchment situated in the upstream of Hutuo River	Mixed watershed	2736	Semi-arid region	372–1746	T: N.M C.P: N.M M: SWAT	23.89	76.11	(Fan et al. 2010)

(continued)

Table A.1 (continued)

Sl. No.	Country and study period	Study area	Type of watershed	Spatial scale (Km^2)	Climatic region	Elevation (m)	Method	Results (%) (Runoff decreased)		References
								CV	HA	
30	China 1961–2006	Hun and Tai River basin	Agricultural watershed	25,364	Temperate continental monsoon climate	0–2744	T: mk test C.P: mk rank statistic & Yt test M: HS Model	Hun: 43.3 Tai: 43.5	Hun: 56.7 Tai: 56.5	(Zhang et al. 2011)
31	China 1956–2000	9 watersheds of the upper catchment of the Yellow River Basin (UYRB)	Mixed watershed	52,500	Semi-arid and semi-humid region	1400–6300	T: N.M C.P: Pettitt test M: Budyko-type (Decomposition method)	5.1–55.5	44.5–94.9	(Wang and Hejazi 2011)
32	China 1956–2000	11 catchments of Loess Plateau	Mixed watershed	113,000	Arid and semiarid continental monsoon climate	1200–1600	T: N.M C.P: Pettitt test M: Budyko-type (Decomposition method)	19.5–71.5	28.5–80.5	(Wang and Hejazi 2011)
33	China 1961–2001	Chao River and Bai River Catchments, Chaobai River of Hai River basin	Mixed watershed	13,846	Temperate continental monsoon climate	50–2262	T: N.M C.P: Pettitt test M: Budyko-type (Decomposition method)	Chao:35.3 Bai: 47.2	Chao:64.7 Bai:52.8	(Wang and Hejazi 2011)

(continued)

Table A.1 (continued)

Sl. No.	Country and study period	Study area	Type of watershed	Spatial scale (Km^2)	Climatic region	Elevation (m)	Method	Results (%) (Runoff decreased)		References
								CV	HA	
34	China 1961–1997	Wuding River basin, Loess Plateau	Mixed watershed	30,261	Semi-arid region	600–1800	T: N.M C.P: Pettitt test M: Budyko-type (Decomposition method)	37.2	62.8	(Wang and Hejazi 2011)
35	China 1964–2008	Semiarid Loahahe basin	Agricultural watershed	18,112	Temperate semi-arid continental monsoon climate	427–2054	T: mk test C.P: Pettitt test M: MR mode; HS model & HMSM	07–11	89–93	(Jiang et al. 2011)
36	Asia 1948–2000	Monsoon Asia	Mixed watershed	–	Monsson climate	Varies	T: N.M C.P: N.M M: DLEM	Less	Much larger	(Tian et al. 2011)
37	China 1956–1996	Wudinghe River basin	Mixed watershed	29,662	Semi-arid region	100–1181	T: N.M C.P: N.M M: HM & SMR	Lesser	Greater	(Xu 2011)
38	China 1964–2008	Laohahe River basin	Agricultural watershed	18,112	Temperate semi-arid continental monsoon climate	427–2054	T: mk test C.P: Pettitt test & PRDCC method M: MR; HS; VIC-3L	89–93	07–11	(Jiang et al. 2011)

(continued)

Table A.1 (continued)

Sl. No.	Country and study period	Study area	Type of watershed	Spatial scale (Km^2)	Climatic region	Elevation (m)	Method	Results (%) (Runoff decreased)		References
								CV	HA	
39	China 1950–2009	Yangtze river basin	Agricultural watershed	7,52,443	Arid to semi-arid region	120–3496	T: N.M C.P: C.A M: SCRCQ	10	90	(Wang et al. 2012a, b)
40	Argentina 1931–2004	Sali-Dulce Basin of Laguna Mar Chiquita Catchment	Agricultural watershed	23,810	Subtropical climate	222–5400	T: SRCC C.P: N.M M: SWAT	More	Less	(Troin et al. 2012)
41	China 1956–2010	Nenjiang River Basin	Forest watershed	297,000	Semi-arid region	76–1740	T: KRCT C.P: Pettitt test M: DCC	69.6–80.3	19.7–30.4	(Dong et al. 2012)
42	China 1960–2008	Tanghe river catchment	Agricultural watershed	3465	Semi-arid region	5–2749	T: mk test C.P: moving t-test M: CE & HMM	38–40	60–62	(Hu et al. 2012)
43	China 1961–2008	Da River Basin of Red River	Forest watershed	55,000	Humid tropical monsoonal region	0–3318	T: mk test C.P: pettitt test M: SBM & MSM	40	60	(Wang et al. 2012a, b)

(continued)

Table A.1 (continued)

Sl. No.	Country and study period	Study area	Type of watershed	Spatial scale (Km^2)	Climatic region	Elevation (m)	Method	Results (%) (Runoff decreased)		References
								CV	HA	
44	China 1957–2000	Upstream of Taolinkou Catchment, Hai River basin (HRB)	Mixed watershed	5060	Semi-humid and semi-arid region	500–2000	T: mk test C.P: mk test M: VIC model	42	59	(Bao et al. 2012)
45	China 1951–2004	Upstream of Guantai, Hai River basin	Agricultural watershed	17,800	Semi-humid and semi-arid region	500–2000	T: mk test C.P: mk test M: VIC model	26	74	(Bao et al. 2012)
46	China 1954–2009	Pearl River basin	Mixed watershed	450,000	Subtropical and tropical monsoon climate	500–2860	T: mk tset C.P: N.M M: LR	18 decreased		(Wu et al. 2012)
47	China 1956–2005	Huifa River basin	Mixed watershed	12,411	Sub-humid area	172–1391	T: N.M C.P: N.M M: SWAT	36.7–45.2 decreased	32.9–67.8 decreased	(Zhang et al. 2012)
48	Australia 1951–2007	Three catchments, Australia	Mixed watersheds	760–1174	Temperate climate	3–440	T: N.M C.P: N.M M: HS model	41–51 decreased	49–59 decreased	(Li et al. 2012)

(continued)

Table A.1 (continued)

Sl. No.	Country and study period	Study area	Type of watershed	Spatial scale (Km^2)	Climatic region	Elevation (m)	Method	Results (%) (Runoff decreased)		References
								CV	HA	
49	China 1957–2000	Hutuo River Basin of Huaihe River Basin	Mixed watershed	15,900	Sem-arid region	165–2960	T: mk test C.P: PRDCC M: HMM & LR	61.9–69.9	30.1–38.1	(Peng et al. 2013)
50	China 1957–2009	Zhang River basin	Mixed watershed	17,800	Temperate continental semi-humid and semi-arid monsoon zone	176–2100	T: mk test C.P: mks test M: HMM & HS	Wet season: 43 Dry season: 72	Wet season: 57 Dry season: 28	(Zeng et al. 2013)
51	China 1986–1998	Bai River basin of Hai River basin	Agricultural watershed	9228	Continental monsoon climate zone	50–2262	T: mk test C.P: MAFU M: Budyko-type & SIMHYD	37.5	62.5	(Zhan et al. 2013)
52	China 1960–2009	Kaidu River Basin	Grassland watershed	18,827	Arid region	1042–4796	T: mk test C.P: MKS test M: HS model	90.5	9.5	(Chen et al. 2013)
53	China 1955–1979	Kuyehe River catchment of middle Yellow River	Agricultural watershed	8645	Arid to semiarid climate	500–2000	T: N.M C.P: SCA M: VIC	32	68	(Wang et al. 2013a, b)

(continued)

Table A.1 (continued)

Sl. No.	Country and study period	Study area	Type of watershed	Spatial scale (Km^2)	Climatic region	Elevation (m)	Method	Results (%) (Runoff decreased)		References
								CV	HA	
54	China 1980–1995	Qingyi River watershed	Forest watershed	13,263	Subtropical humid zone	398–5298	T:mk test C.P: N.M M: VIC model	2.4 decreased	0.2 decreased	(Liu et al. 2013)
55	China 1963–2008	Chaohe watershed of Miyun reservoir	Agricultural watershed	4854	Temperate continental	159–2218	T: N.M C.P: N.M M: EA; DEBA & MIKESHE	45–78	22–55	(Wang et al. 2013a, b)
56	China 1960–2006	Kuye River basin of the Middle Yellow River	Agricultural watershed	8706	Continental monsoon climate	500–2000	T: mk test C.P: N.M M: Budyko-type (CE model) & SIMHYD	39.3–40.9	59.1–60.7	(Liang et al. 2013)
57	China 1957–2000	Luanhe, Chaohe, Hutuo and Zhanghe of Haihe river basin	Agricultural watershed	318,000	Temperate continental monsoon region	100–3059	T: mk tset C.P: N.M M: HM, HS & CE models	Lesser (except Hutuo)	> 50 (except Hutuo)	(Wang et al. 2013a, b)
58	China 1957–2000	Upstream of Luanxian Catchment, Luan River of Hai River basin	Agricultural watershed	44,100	Temperate continental monsoon type	0.4–3047	T: mk test C.P: Pettitt test & PRDCC M: HM, HS & CE models	33–43	57–67	(Wang et al. 2013a, b)

(continued)

Table A.1 (continued)

Sl. No.	Country and study period	Study area	Type of watershed	Spatial scale (Km^2)	Climatic region	Elevation (m)	Method	Results (%) (Runoff decreased)		References
								CV	HA	
59	China 1956–2005	Upstream of Panjiakau Catchment, Luan River of Hai River basin	Agricultural watershed	33,700	Temperate continental monsoon type	144–2228	T: mk test C.P: GP time vs Ruoff M: GBHM & CE models	45.1–47.2	54.9–52.8	(Xu et al. 2013)
60	China 1957–2008	Jinghe River Basin	Agricultural watershed	2150	Continental arid and semi-arid zones	423–2922	T: N.M C.P: SC method M: SWAT	85.7	14.3	(Dong et al. 2013)
61	China 1957–2000	Chao River Catchment, Chaobai River of Hai River basin	Agricultural watershed	4701	Temperate continental monsoon type	0.4–3047	T: mk test C.P: Pettitt test & PRDCC M: HM, HS & CE models	34–46	54–66	(Wang et al. 2013a, b)
62	China 1957–2000	Upstream of Guanting, Yongding River of Hai River basin	Agricultural watershed	47,016	Semi-arid region	0–3059	T: N.M C.P: N.M M: SWAT	65	35	(Zhang et al. 2013)

(continued)

Table A.1 (continued)

Sl. No.	Country and study period	Study area	Type of watershed	Spatial scale (Km^2)	Climatic region	Elevation (m)	Method	Results (%) (Runoff decreased)		References
								CV	HA	
63	China 1956–2000	Upstream of Guantai, Zhang River of Hai River basin	Agricultural watershed	17,800	Temperate continental monsoon type	0.4–3047	T: mk test C.P: Pettitt test & PRDCC M: HM, HS & CE models	31–36	64–69	(Wang et al. 2013a, b)
64	China 1961–2000	Upper and middle reaches of the Taoer River	Forest watershed	27,633	Cold and long winters and hot and short summers	150–1709	T: mk test C.P: moving t-test M: HM & Budyko-type	45	55	(Li et al. 2013)
65	China 1970–2000	Upper Hanjiang River Basin (UHR) of the Yangtze River	Agricultural watershed	95,200	Semi-humid climate	88–3500	T: mk test C.P:smk test M: Budyko-type (CE & DM) & DHM	Greater	Lesser	(Sun et al. 2014)
66	China 1956–2005	12 catchments of Luan River in Hai River basin	Agricultural watershed	1086–171,000	Semi-arid region	0–3059	T: mk test C.P: Pettitt test M: Budyko-type (CE method)	19	81	(Xu et al. 2014a, b)

(continued)

Table A.1 (continued)

Sl. No.	Country and study period	Study area	Type of watershed	Spatial scale (Km^2)	Climatic region	Elevation (m)	Method	Results (%) (Runoff decreased)		References
								CV	HA	
67	China 1956–2005	12 catchments of Luan River in Hai River basin	Agricultural watershed	1086–171,000	Semi-arid region	0–3059	T: mk test C.P: N.M M: Budyko-type (modified CE method)	33–34	66–67	(Yang et al. 2014)
68	China 1956–2005	4 Catchments, Chaobai River of Hai River basin	Agricultural watershed	1600–4266	Semi-arid region	0–3059	T: mk test C.P: Pettitt test M: Budyko-type (CE method)	19	81	(Xu et al. 2014a, b)
69	China 1956–2005	4 Catchments, Chaobai River of Hai River basin	Agricultural watershed	1600–4266	Semi-arid region	0–3059	T: mk test C.P: N.M M: Budyko-type (modified CE method)	33–34	66–67	(Yang et al. 2014)
70	China 1956–2005	5 Catchments, Yongding River of Hai River basin	Mixed watershed	47,016	Semi-arid region	0–3059	T: mk test C.P: Pettitt test M: Budyko-type (CE method)	19	81	(Xu et al. 2014a, b)

(continued)

Table A.1 (continued)

Sl. No.	Country and study period	Study area	Type of watershed	Spatial scale (Km^2)	Climatic region	Elevation (m)	Method	Results (%) (Runoff decreased)		References
								CV	HA	
71	China 1956–2005	5 Catchments, Yongding River of Hai River basin	Mixed watershed	47,016	Semi-arid region	0–3059	T: mk test C.P: N.M M: Budyko-type (modified CE method)	33–34	66–67	(Yang et al. 2014)
72	China 1956–2005	7 Catchments, Daqing River of Hai River basin	Mixed watershed	1611–11,936	Semi-arid region	0–3059	T: mk test C.P: Pettitt test M: Budyko-type (CE method)	19	81	(Xu et al. 2014a, b)
73	China 1956–2005	7 Catchments, Daqing River of Hai River basin	Mixed watershed	1611–11,936	Semi-arid region	0–3059	T: mk test C.P: N.M M: Budyko-type (modified CE method)	33–34	66–67	(Yang et al. 2014)
74	China 1956–2005	3 Catchments, Ziya River of Hai River basin	Agricultural watershed	2521–6420	Semi-arid region	0–3059	T: mk test C.P: Pettitt test M: Budyko-type (CE method)	19	81	(Xu et al. 2014a, b)

(continued)

Table A.1 (continued)

Sl. No.	Country and study period	Study area	Type of watershed	Spatial scale (Km^2)	Climatic region	Elevation (m)	Method	Results (%) (Runoff decreased)		References
								CV	HA	
75	China 1956–2005	3 Catchments, Ziya River of Hai River basin	Agricultural watershed	2521–6420	Semi-arid region	0–3059	T: mk test C.P: N.M M: Budyko-type (modified CE method)	33–34	66–67	(Yang et al. 2014)
76	China 1956–2005	2 Catchments, Zhang River of Hai River basin	Mixed watershed	5060–11,250	Semi-arid region	0–3059	T: mk test C.P: Pettitt test M: Budyko-type (CE method)	19	81	(Xu et al. 2014a, b)
77	China 1956–2005	2 Catchments, Zhang River of Hai River basin	Mixed watershed	5060–11,250	Semi-arid region	0–3059	T: mk test C.P: N.M M: Budyko-type (modified CE method)	33–34	66–67	(Yang et al. 2014)
78	China 1950s-2010	Middle reaches of the Yellow River basin	Agricultural watershed	362,000	Semi-arid climate zone	500–2000	T: mk test C.P: Pettitt test M: Budyko-type (WB model)	Varies across basin	Varies across basin	(Zhao et al. 2014)

(continued)

Table A.1 (continued)

Sl. No.	Country and study period	Study area	Type of watershed	Spatial scale (Km^2)	Climatic region	Elevation (m)	Method	Results (%) (Runoff decreased)		References
								CV	HA	
79	USA 1950–2010	Four states of USA namely New York, Indiana, Arizona and Georgia	Mixed watershed	> 100,000	Humid continental climate conditions to dry climate	0–2281	T: N.M C.P: N.M M: LR, HS, AB & Budyko-type	Lesser	Greater	(Ahn and Merwade 2014)
80	China 1975–2009	A small catchment (Xitiaoxi river basin) in Loess plateau	Mixed watershed	1320	Subtropical monsoon climate	3–1564	T: mk test C.P: MKS test and moving t-test M: WB model	26	74	(Zhang et al. 2014a, b)
81	China 1972–2007	Upper reaches of Weihe River	Agricultural watershed	30,661	Semi-arid continental climate	647–3635	T: mk test C.P: Pettitt test & PRDCC M: HMSM & HS	29–41 23–34	59–71 66–77	(Guo et al. 2014)
82	China 1960–2009	Several catchments of Wei River Basin (WRB)	Agricultural watershed	135,000	Temperate continental monsoon climate	300–3865	T: mk test C.P: Pettitt test M: HS & WB models	29–65 decreased	31–65 decreased	(Zuo et al. 2014)

(continued)

Table A.1 (continued)

Sl. No.	Country and study period	Study area	Type of watershed	Spatial scale (Km^2)	Climatic region	Elevation (m)	Method	Results (%) (Runoff decreased)		References
								CV	HA	
83	China 1975–2009	Xitiaoxi river basin	Agricultural watershed	1320	Sub-tropical climate	3–1564	T: mk test C.P: moving t-test & mks test M: SWAT	42.8	57.2	(Zhang et al. 2014a, b)
84	China 1958–2009	Luan River Basin	Agricultural watershed	44,750	Temperate continental monsoon climate	0–2205	T: mk test C.P: mk test & MAFU M: SWAT, SIMHYD, Budyko-type, DMC methods	28.3–46.8	53.2–71.7	(Zeng et al. 2015)
85	Ethiopia 1985–2006	Central Rift Valley (CRV) basin	Agricultural watershed	10,185	Semi-arid climate	1525–4201	T: mk tset C.P: N.M M: SWAT	30	70	(Seyoum et al. 2015)
86	China 1956-2000s	Wei River Basin (WRB)	Agricultural watershed	135,000	Temperate continental monsoon	300–3865	T: N.M C.P: N.M M: VIC	90	10	(Chang et al. 2015)
87	China 1961–2009	14 catchments in Loess Plateau	Agricultural watershed	752,444	Arid and semiarid continental monsoon climate	78–4950	T: N.M C.P: N.M M: Budyko-type (CE & DM)	38	62	(Liang et al. 2015)

(continued)

Table A.1 (continued)

Sl. No.	Country and study period	Study area	Type of watershed	Spatial scale (Km^2)	Climatic region	Elevation (m)	Method	Results (%) (Runoff decreased)		References
								CV	HA	
88	China 1960–2009	Weihe River basin, China	Agricultural watershed	134,800	Semi-humid region to an arid region	647–3635	T: N.M C.P: N.M M: Budyko-type (DM)	Greater	Lesser	(Jiang et al. 2015)
89	Tanzania 1972–2010	Ngerengere River basin	Agricultural watershed	2780	Tropical climate	98–2260	T: mk test C.P: FDC M: SWAT	Decreased	Decreased	(Natkhin et al. 2015)
90	China 1956–2010	Four sub-watersheds in Luanhe watershed in	Mixed watersheds	44,600	Semi-arid region	Feb-05	T: N.M C.P: N.M M: SWAT	58–65 decreased	35–42 decreased	(Li et al. 2015)
91	China 1954–2012	Huangfuchuan River basin (HFCRB) of the Loess Plateau are	Agricultural watershed	3246	Semi-arid climate	1000–3000	T: mk test C.P: Pettitt test M: SWAT	53.7 decreased	25.3 decreased	(Zuo et al. 2015)
92	West Africa 1976–2011	Black Volta basin	Agricultural watershed	130,400	Semi-humid to humid	60–762	T: mk test C.P: N.M M: SWAT	Not significant	Significant	(Akpoti et al. 2016)

(continued)

Table A.1 (continued)

Sl. No.	Country and study period	Study area	Type of watershed	Spatial scale (Km^2)	Climatic region	Elevation (m)	Method	Results (%) (Runoff decreased)		References
								CV	HA	
93	China 1961–2009	14 catchments of the Loess Platea	Mixed watershed	49,000	Semi-arid climate	100–5000	T: LR C.P: MTW M: PCM	44	56	(Feng et al. 2016)
94	China 1960–2010	Wei River basin	Agricultural watershed	134,800	Semi-humid region to an arid region	1100–3927	T: mk test C.P: mk test M: SWAT	70 in 1990s Lesser in 2000s	30 in 1990s Greater in 2000s	(Li et al. 2016)
95	China and Kyrgyzstan 1960–2011	Aksu River basin	Agricultural watershed	43,000	Arid region	1087–7126	T: mk test C.P: AAM M: SWAT	92.28	7.72	(Meng et al. 2016)
96	China 1961–2011	Yihe River basin	Agricultural watershed	2856	Semi humid basin	368–713	T: mk test C.P: N.M M: LR & CE	56.38–67.68	33.62–32.32	(Saifullah et al. 2016)
97	China 1961–2010	Jing River catchment of the Yellow River	Agricultural watershed	45,421	Sem-arid region	416–2908	T: N.M C.P: N.M M: Budyko-type	Greater	Lesser	(Ning et al. 2016)
98	China 1970–2010	Weihe River Basin	Agricultural watershed	135,000	Arid to sub-humid areas	319–3917	T: mk test C.P: N.M M: SWAT	56.63 mm/ year decreased	28.07 mm/ year decreased	(Zhao et al. 2016)

(continued)

Table A.1 (continued)

Sl. No.	Country and study period	Study area	Type of watershed	Spatial scale (Km^2)	Climatic region	Elevation (m)	Method	Results (%) (Runoff decreased)		References
								CV	HA	
99	China 1961–2000	Hutuo River Basin (HRB)	Agricultural watershed	15,900	Semi-arid region	127–3059	T: mk test C.P: Pettitt test M: SWAT	45.30 decreased	2.06 decreased	(Omer et al. 2016)
100	South Korea 1974–2014	Soyang Dam upper basin and the Seom River basin of Han River	Forest watershed	2703 1491	Humid region	1–1700	T: mk test C.P: Pettitt test, DMC & BCP M: SWAT & Budyko-type	Soyang Dam: greater Seom River: lesser	Soyang Dam: lesser Seom River: greater	(Lee and Kim 2017)
101	China 1986–2012	Upper Sang-kan (USK) river basin of Haihe river basin, Loess Plateau	Agricultural watershed	3430	Arid to semi-arid climate	1027–2603	T: N.M C.P: N.M M: SWAT	39.1	63.1–64.8	(Zhang et al. 2017a, b)
102	China 1951–2016	Three Gorges Reservoir, upstream Yangtze River	Agricultural watershed	1,000,000	Sub-tropical region	51–7143	T: LR & mk test C.P: Pettitt test & SCM M: SWAT	70	30	(Zhang et al. 2017a, b)
103	China 1960–2015	Tarim River basin	Mixed watershed	1,020,000	Extremely arid area	767–8531	T: mk test C.P: Pettit test M: DMC and Budyko-type	Greater	Lesser	(Xue et al. 2017)

(continued)

Table A.1 (continued)

Sl. No.	Country and study period	Study area	Type of watershed	Spatial scale (Km^2)	Climatic region	Elevation (m)	Method	Results (%) (Runoff decreased)		References
								CV	HA	
104	China 1954–2013	Gushan river basin	Agricultural watershed	1263	Arid to semi-arid zone	100–4000	T: LR C.P: SCM M: SIMHYD	38.5	61.5	(Wang et al. 2017)
105	India 1983–2008	Ungauged river basin (Sirsa river of Satluj River), NW Himalaya	Mixed watershed	680	Sub-tropical monsoon climate	250–1900	T: mk test C.P: N.M M: SWAT	More	Less	(Verma and Dash 2018)
106	China 1969–2012	Miyun Reservoir basin	Mixed watershed	15,350	Continental monsoon climate	88–2289	T: N.M C.P: N.M M: SWAT	45.7–64.9	35.1–54.3	(Yan et al. 2018)
107	China 1972–2011	Yanhe River Basin, middle reaches of the Yellow River	Agricultural watershed	7485	Warm temperate continental monsoonal climate	500–1780	T: LR C.P: Pettitt test M: SWAT	55.8	44.2	(Wu et al. 2018)
108	China 1957–2010	Luanhe river basin	Agricultural watershed	44,750	Sub-arid region	Feb-05	T: mk test and MAM C.P: OC & moving t-test M: SWAT	55.86–77.14	22.86–43.14	(Zhang et al. 2018a, b)

(continued)

Table A.1 (continued)

Sl. No.	Country and study period	Study area	Type of watershed	Spatial scale (Km^2)	Climatic region	Elevation (m)	Method	Results (%) (Runoff decreased)		References
								CV	HA	
109	China 1960–2012	Fuping hydrological station basin	Mixed watershed	2200	Temperate continental monsoon semi-humid and semi-arid climate	250–2784	T: mk test C.P: Pettitt test M: SWAT	37.5	62.5	(Wang et al. 2018)
110	China 1960–2005	Dongjiang River Basin of Pearl River basin	Agricultural watershed	27,040	Subtropical climate	1–1500	T: LR C.P: mk test M: LR, SWAT & CE	58	42	(Zhou et al. 2018)
111	China 1969–2015	Xihe River Basin	Mixed watershed	1047	Semi-humid and semi-arid region	75–1157	T: LR C.P: Pettitt test M: SWAT	26.54 decreased	126.54 decreased	(Guo et al. 2018)
112	India 1990–2011	Ungauged catchment (32 catchments from Eastern and Southern India, spreading over 9 river basins)	Mixed watersheds	–	Monsson climate	1–1264	T: mk test C.P: N.M M: SWAT	Higher	Smaller	(Swain and Patra 2019)

(continued)

Table A.1 (continued)

Sl. No.	Country and study period	Study area	Type of watershed	Spatial scale (Km^2)	Climatic region	Elevation (m)	Method	Results (%) (Runoff decreased)		References
								CV	HA	
113	India 1973–2013	Tapi river basin	Agricultural watershed	65,145	Monsoonal climate	0–762	T: mmk test & SR C.P: Pettit, M-W U-test & SS M: UC method	15	85	(Sharma et al. 2019)
114	USA 1980–2017	Urban watershed, Little Eagle Creek (LEC) watershed	Urban watershed	74.5	Mediterranean climate	208–275	T: mk test C.P: mk test M: SWAT	Greater	Lesser	(Aboelnour et al. 2019)
115	China 1978–2016	Intensive Phosphate Mined Huangbaihe River basin	Agricultural watershed	554	Continental monsoon climate	221–1781	T: mk test C.P: AAM M: SWAT & SCRAQ	Greater during 1985–2005 60.5 74.7	Greater during 2006–2016 79.3 86.1	(Bo et al. 2019)
116	Iran 1975–2010	Haraz River basin	Mixed watershed	4014	Semi-arid region	300–5600	T: sequential mk test C.P: N.M M: HS, DMC & SWAT	29.86–40.19	59.81–70.14	(Pirnia et al. 2019)

(continued)

Table A.1 (continued)

Sl. No.	Country and study period	Study area	Type of watershed	Spatial scale (Km^2)	Climatic region	Elevation (m)	Method	Results (%) (Runoff decreased)		References
								CV	HA	
117	China 1956–2005	Wei River basin of Yellow river	Agricultural watershed	135,000	Semi-humid and semi-arid region	304–3836	T: N.M C.P: N.M M: WEP-L	75.2	24.8	(Liu et al. 2019)
118	China 1970–2016	Weihe river basin (WRB) of the Yellow River	Agricultural watershed	134,800	Transition zone of arid and humid regions	304–3836	T: mk test & LR C.P: mk test M: SWAT	44.3–49.7	50.3–55.7	(Li et al. 2019a, b)
119	China 1980–2008	Upper Heihe River Basin	Grassland watershed	10,000	Arid and semiarid climate	1600–4800	T: mk test & CAM C.P: mk test M: SWAT	87.15	12.85	(Shang et al. 2019)
120	Ethiopia 1980–2013	Ketar and Meki watersheds of Lake Ziway basin	Agricultural watershed	7311	Monsoon tropical climate	1624–4189	T: mk test C.P: N.M M: SWAT	Greater	Lesser	(Musie et al. 2020)
121	China 1951–2013	Yihe River basin of Huaihe River	Agricultural watershed	17,000	Semi-humid continental climate	200–2159	T: mk test & EEMD C.P: N.M M: SWAT & WRF	More	Less	(Li et al. 2020a, b)

(continued)

Table A.1 (continued)

Sl. No.	Country and study period	Study area	Type of watershed	Spatial scale (Km^2)	Climatic region	Elevation (m)	Method	Results (%) (Runoff decreased)		References
								CV	HA	
122	China 2006–2018	Guishui river basin	Agricultural watershed	626	Continental monsoon climate	473–2212	T: N.M C.P: N.M M: SWAT	60	40	(Wang et al. 2020)
123	China 1980–2012	Upper Hailar Basin	Grassland watershed	54,500	Temperate continental monsoon climate	598–1706	T: mk test C.P: sliding t-test M: SWAT	83.58	16.42	(Yan et al. 2020)
124	China 1956–2015	3 typical and small mountainous areas of Haihe River Basin (HRB)	Mixed watershed	318,000	Continental monsoon climate	2–2836	T: N.M C.P: DMC M: SWAT	32.48–38.17	61.83–67.52	(Jiao et al. 2020)
125	China 1981–2014	Min–Tuo River Basin	Agricultural watershed	135,200	Sub-tropical climate	240–7556	T: mk test C.P: sliding t-test, CA method & mk test M: SWAT & Budyko-type	93.26	6.74	(Hu et al. 2020)

(continued)

Table A.1 (continued)

Sl. No.	Country and study period	Study area	Type of watershed	Spatial scale (Km^2)	Climatic region	Elevation (m)	Method	Results (%) (Runoff decreased)		References
								CV	HA	
126	China 1976–2000	Upstream of the Lancang River Basin	Mixed watershed	90,000	Sem-arid region	1283–6456	T: N.M C.P: N.M M: SWAT	Greater	Lesser	(Ma et al. 2020a, b)
127	China 1971–2016	Zhou River Basin (ZRB)	Agricultural watershed	774	Temperate continental monsoon climate	1221–1793	T: mk test C.P: Pettitt test M: SWAT	Decreased	Decreased	(Sun et al. 2020)
128	Afghanistan 1990–2011	Middle Helmand Basin	Mixed watershed	90,000	Arid to semi-arid climate	400–5000	T: mk test C.P: Pettitt test M: SWAT	Small impact	810 MCM/ year decreased	(Hajihosseini et al. 2020)
129	China 1985–2014	Min River watershed of Yangtze River basin	Agriculture watershed	22,500	Arid to semi-arid climate	653–6054	T: N.M C.P: N.M M: SWAT	58.53 decreased	2.38 decreased	(Ma et al. 2020a, b)
130	China 1981–2016	Soft sandstone area of the middle Yellow River basin	Agriculture watershed	16,700	Arid to semi-arid climate	773–1625	T: mk test C.P: Pettitt test M: Budyko-type	8	92	(Li et al. 2020a, b)

(continued)

Table A.1 (continued)

Sl. No.	Country and study period	Study area	Type of watershed	Spatial scale (Km^2)	Climatic region	Elevation (m)	Method	Results (%) (Runoff decreased)		References
								CV	HA	
131	China 1982–2012	Yanhe watershed from	Agriculture watershed	7689	Arid region	495–1793	T: mk test C.P: Pettitt test M: SWAT & statistical method	Greater	Lesser	(Kang et al. 2020)
132	China 1964–2003	Zhenjiangguan watershed	Forested watershed	4486	Humid climate	2445–5360	T: CSDC C.P: mk test M: BPCC & RDA	83	17	(Zhang et al. 2020a, b)
133	China 1951–2013	Upper Yangtze River basin	Agriculture watershed	1,000,000	Sem-humid type	938–3272	T: mk test C.P: Pettitt test & SCM M: SWAT, statistical model & conceptual analysis	70	30	(Zhang et al. 2021)
134	China 1988–2012	Yangtze River basin	Agriculture watershed	1,690,000	Subtropical and temperate climate	0–7046	T: N.M C.P: N.M M: SWAT	Greater	Lesser	(Nie et al. 2021)

(continued)

Table A.1 (continued)

Sl. No.	Country and study period	Study area	Type of watershed	Spatial scale (Km^2)	Climatic region	Elevation (m)	Method	Results (%) (Runoff decreased)		References
								CV	HA	
135	China 1963–2015	Lancang River	Mixed watershed	160,000	Semi-arid region	547–5871	T: mk test C.P: mk break point test M: SWAT, Budyko-type & GMSWU	37.2–42.6	57.4–62.85	(Tang et al. 2021)
136	China 1955–2017	Zhangfang, Weishui and Guantai basins, Taihang mountain region	Agriculture watershed	134,900	Warm and humid type	15–3004	T: mk test C.P: mk test M: SWAT	31.61–45.81	54.19–68.39	(Wang et al. 2021)
137	Uganda 1997–2019	River Rwizi catchment	Agriculture watershed	8555	Semi-humid	1166–2171	T: N.M C.P: N.M M: SWAT	73	27	(Onyutha et al. 2021)
138	Iran 1985–2014	Zilbier River basin	Grassland watershed	2527	Arid and semi-arid	938–3572	T:mk test C.P: N.M M: SWAT	83.6	16.4	(Andaryani et al. 2021)

Source Authors

Table A.2 Inventory of literature which shows increasing/decreasing runoff tendency

Sl. No.	Country & study period	Study area	Type of watershed	Spatial scale (Km^2)	Climatic region	Elevation (m)	Method	Results (%)		References
								CV	HA	
1	USA 1940–1990	Four river basins situated in Indiana and Chicago	Mixed watersheds	895–5960	N.M	N.M	T: SLRM & mk test C.P: N.M M: Paired Catchment	Increased	Increased	(Changnon and Demissie 1996)
2	Canada 1934–1998	Great Slave Lake	Mixed watershed	28,568	Sub-Arctic Climate	156 m (mean)	T: MWRST C.P: N.M M: WB & LR	7–13 cm/ year increased	11–17 cm/ year decreased	(Gibson et al. 2006)
3	USA 1941–2000	Coon Creek and Kickapoo River Watersheds	Agricultural watershed	4554	Humid continental climate	183–518	T: N.M C.P: N.M M: ARIMA	Increased	Increased	(Juckem et al. 2008)
4	China 1965–2005	Kejie Watershed in the Salween River Basin in Yunnan	Agricultural watershed	1755	Sub-tropical climate	963–3076	T: mk test C.P: PLM M: LR	1 increased	3.7 decreased	(Ma et al. 2009)
5	USA 1925–2003	Four Large agricultural watersheds of the Midwest USA	Agricultural watersheds	3240–8912	Humid continental climate	150–460	T: N.M C.P: N.M M: SRRC Model	Increased in one watershed	Increased in one watershed	(Tomer and Schilling 2009)
6	China 1965–2005	Kejie Watershed	Agricultural watershed	1755	Sub-tropical climate	963–3076	T: mk test C.P: MW test M: SWAT	1 increased	3.7 decreased	(Ma et al. 2009)
7	Thailand 1990–2004	Yom watershed	Mixed watershed	25,180	Monsoon	20–360	T: N.M C.P: N.M M: LR & MIKE 11	Increase	Increase	(Petchprayoon et al. 2010)
8	USA 1948–2003	413 watersheds in the contiguous United States (nonurban)	Mixed watersheds	–	Humid continental to cold semi-arid type	460–1500	T: N.M C.P: Pettitt test M: Budyko-type (Decomposition method)	Increased	Increased in Midwest USA and decreased in high plains	(Wang and Hejazi 2011)

(continued)

Table A.2 (continued)

Sl. No.	Country & study period	Study area	Type of watershed	Spatial scale (Km^2)	Climatic region	Elevation (m)	Method	Results (%)		References
								CV	HA	
9	Australia 1920–2007	Comet River & Upper Burdekin catchments	Grassland watershed	16,440 and 40,000	Dry sub-humid	150–1223 and 350–900	T: mk test C.P: FDC M: Budko-type & TSF	Decreased	Increased	(Peña-Arancibia et al. 2012)
10	China 1986–2005	Xixian Basin of Huaihe River	Agricultural watershed	10,191	Subtropical region to warm temperate zone	10–100	T: N.M C.P: N.M M: SWAT	Increased		(Shi et al. 2013a, b)
11	China 1964–2005	Xixian Basin of Huaihe River	Agricultural watershed	10,191	Sub-tropical to temperate zone	10–100	T: mk test C.P: MW test M: SWAT	4.6 increased	2.1 decreased	(Peng et al. 2013)
12	China 1960–2007	Poyang Lake catchment, middle reaches of the Yangtze River	Agricultural watershed	162,225	Subtropical wet climate zone	32–2200	T: mk test C.P: SMKT M: HS	105–212.1 increased	5–112 decreased	(Ye et al. 2013)
13	USA 1934–2005	38 basins of South Atlantic	Mixed watersheds	852–21,714	Humid-subtropical climate	Sea level-2037	T: N.M C.P: N.M M: Budyko-type	14 increased	> 27 increased	(Patterson et al. 2013)
14	West Africa 1950–2009	Sahelian part of the Niger River basin	Grassland watersheds	297,400	Semi-arid region	157–1693	T: N.M C.P: N.M M: SWIM	Increased	Increased	(Aich et al. 2015)
15	U.S.A 1970–2009	Neches River Basin	Forest watershed	2221	Humid subtropical climate	0–152	T: N.M C.P: N.M M: SWAT	Increased	Increased	(Heo et al. 2015a, b)
16	Indonesia 1984–2006	Upper Brantas river basin	Agricultural watershed	1912	Tropical monsoon climate	240–3165	T: mk test C.P: N.M M: SWAT	19.07 decreased	20.14 increased	(Setyorini et al. 2017)
17	India 1990–2008	Harangi river basin	Agricultural watershed	420	Tropical monsoon climate	900–1800	T: mk test C.P: N.M M: SWAT	Decreased	0.77 increased	(Anil and Ramesh 2017)

(continued)

Table A.2 (continued)

Sl. No.	Country & study period	Study area	Type of watershed	Spatial scale (Km^2)	Climatic region	Elevation (m)	Method	Results (%)		References
								CV	HA	
18	Italy 1954–2010	Elsa river basin, a rural hilly basin in Italy	Mixed watershed	931	European Mediterranean area	46–300	T: N.M C.P: N.M M: SWAT	Increased	Increased	(Napoli et al. 2017)
19	China 1970s-2000s	Jinghe River basin	Agricultural watershed	> 45,000	Semi-humid to arid region	360–2900	T: N.M C.P: N.M M: SWAT	Increased		(Yin et al. 2017)
20	China 1980–2010	Yingluoxia catchment (YLC) region & Minxian catchment (MXC)	Agricultural watershed	10,009 & 14,912	Semi-arid region	1668–5058 and 2411–4562	T: N.M C.P: N.M M: SWAT	3.2 increased	4.3 decreased	(Yang et al. 2017)
21	China 1995–2014	Upper Heihe River basin	Grassland watershed	10,000	Transitional zone between semiarid and arid areas	1408–5228	T: N.M C.P: N.M M: SWAT & DHSVM	101–102 increased	1–2 decreased	(Zhang et al. 2018a, b)
22	Vietnam 1983–2015	La Buong River Basin	Mixed watershed	479	Tropical monsoon climate	0–2285	T: N.M C.P: N.M M: SWAT	1.6–2.9 decreased	12.9 increased	(Khoi et al. 2019)
23	China 1960–2012	Qingliu River basin, lower Yangtze River	Agricultural watershed	1070	Humid subtropical climate zone	24–339	T: mk test C.P: mk test M: SWAT	16.05 increased	16.87 increased	(Yang et al. 2019)
24	China 1995–2015	Xihe River Basin	Agricultural watershed	1111	Temperate monsoon climate	115–1134	T: N.M C.P: N.M M: SWAT	Increased	Increased	(Lyu et al. 2019)

(continued)

Table A.2 (continued)

Sl. No.	Country & study period	Study area	Type of watershed	Spatial scale (Km^2)	Climatic region	Elevation (m)	Method	Results (%)		References
								CV	HA	
25	China 1990–2010	Zhengshui River Basin, agricultural Watershed	Agricultural watershed	3470	Subtropical monsoon climate	37–1157	T: LR C.P: N.M M: SWAT	100.22 decreased	0.20 increased	(Luo et al. 2020)
26	Mongolia, Russia and China 1980–2013	Amur River Basin, northeastern Asia	Forest watershed	2 M	Harsh continental with monsoon	0–3430	T: mk test & LR C.P: N.M M: SWAT	31.38 decreased	16.17 increased	(Zhou et al. 2020)
27	Sothern Africa 1984–2015	Wamkurumadzi River basin of Shire River in Malawi country	Agricultural watershed	586	Tropical wet and dry Savanna climate	0–2200	T: mk test C.P: N.M M: SWAT	Increased	Increased	(Nkhoma et al. 2021)

Source: Authors

N.B AAM—accumulative anomaly method; AB—annual balance; ARIMA—auto regressive integrated moving average; BCP—Bayesian change point; BP-ANN—back-propagation artificial neural network; BPCC—pollution calculation center; C.A—cumulative anomaly; CAM—cumulative anomaly method; CE—climate elasticity model; CHARM—catchment hydrology and rural management; C.P—change point detection; CSDC—cumulative standard deviation curves; CV—climate variability; DCC—double cumulative curve; DEBA—differential elasticity-based analysis; DHM—dynamic hydrological model; DHSVM—distributed hydrology soil vegetation model; DLEM—dynamic land ecosystem model; DM—decomposition method; DMC—double mass curve; DPBM—distributed physically based model; DTVGM—distributed monthly water balance model; DWBS—dynamic water balance simulation model; EA—ecohydrological approach; EEMD—ensemble empirical mode decomposition; GBHM—grid based hydrological model; GMSWU—global gridded monthly sectoral water use data; GP—graphical plotting; HA—human activities; HS—hydrological sensitivity analysis; H.M—hydrological model; HMM—hydrological modeling method; HMSM—hydrologic model simulation method; KRCT—Kendall's rank correlation test; LFM—linear fitted model; LR—linear regression; M—model/method; MAM—moving average method; MAFU—moving average of FU's formula; MDMC—modified double mass curve method; mk test—Mann-Kendall test, mks—Mann-Kendall-Sneyers test; MR—multi-regression; MSM—model simulation method; MTW—moving time window; MWRST—Mann–Whitney Rank Sum Test; NDIC—normalized differential integral curve; N.M—not mentioned; OC—ordered clustering analysis; PCM—partial correlation method; PLM—piecewise linear model; PMW—Pettitt–Mann–Whitney statistical test; PRDCC—precipitation-runoff double cumulative curve method; PRMS—precipitation runoff modelling system; RDA—redundancy analysis; RHESSys—regional hydro-ecologic simulation system; RM—regression model; SBM—sensitivity-based method; SC—sequential cluster; SCA—sequential cluster analysis change point/critical year detection technique; SCM—sequential clustering method; SCRCQ—slope change ratio of cumulative quantity; SIMHYD—simulation of hydrology for drainage; SLRM—simple linear regression model; SMAR—soil moisture and accounting model; SMKT—sequential Mann-Kendall test; SMR—stepwise multiple regression; S.P—study period; SRCS—Spearman's rank correlation statistics; SRRC—simplified rainfall runoff conceptual model; SRCT—Spearman's rank correlation test; STREAM—spatial tools for river and environmental analysis and management; SWAT—soil and water assessment tool; SWIM—soil and water integrated model; T—trend analysis; TSF—Tomer-schilling framework; TTA—time trend analysis technique; UC—unimpaired catchment; VIC—variable infiltration capacity hydrologic model; VIC-3L—three-layer variable infiltration capacity hydrologic model; WB—water balance; WEP-L—water erosion prediction project long-term hydrological model; WRF—weather research and forecasting model; YM—Yamamoto method; Yt-yest—Yamamoto method based on t-test

References

Aboelnour M, Gitau MW, Engel BA (2019) Hydrologic response in an urban watershed as affected by climate and land-use change. Water (Switzerland) 11(8):1–23. https://doi.org/10.3390/w11081603

Afonso de Oliveira Serrão E, Silva MT, Ferreira TR, Paiva de Ataide LC, Assis dos Santos C, Meiguins de Lima AM, de Paulo Rodrigues da Silva V, de Assis Salviano de Sousa F, Cardoso Gomes DJ (2022) Impacts of land use and land cover changes on hydrological processes and sediment yield determined using the SWAT model. Int J Sediment Res 37(1):54–69. https://doi.org/10.1016/j.ijsrc.2021.04.002

Ahirwar A, Jain MK, Perumal M (2018) Performance of the Xinanjiang Model. In: Jain MK, Singh R, Chatterjee P (eds) Hydrologic modeling. Springer Singapore, pp 715–731. https://doi.org/10.1007/978-981-10-5801-1_49

Ahn KH, Merwade V (2014) Quantifying the relative impact of climate and human activities on streamflow. J Hydrol 515(April):257–266. https://doi.org/10.1016/j.jhydrol.2014.04.062

Aich V, Liersch S, Vetter T, Andersson JCM, Müller EN, Hattermann FF (2015) Climate or land use?Attribution of changes in river flooding in the Sahel zone. Water (Switzerland) 7(6):2796–2820. https://doi.org/10.3390/w7062796

Akpoti K, Antwi EO, Kabo-bah AT (2016) Impacts of rainfall variability, land use and land cover change on stream flow of the Black Volta basin, West Africa. Hydrology 3(3):1–24. https://doi.org/10.3390/hydrology3030026

Alexandersson H (1986) A homogeneity test applied to precipitation data. J Climatol 6(6):661–675. https://doi.org/10.1002/joc.3370060607

Ali R, Kuriqi A, Abubaker S, Kisi O (2019) Long-term trends and seasonality detection of the observed flow in Yangtze River using Mann-Kendall and Sen's innovative trend method. Water 11(9):1855. https://doi.org/10.3390/W11091855

Alice Robson A, Bardossy A, Jones D, Kundzewicz ZW (2000) Statistical methods for testing for change. In: Kundzewicz ZW, Robson A (ed) Detecting trend and other changes in hydrological data, world climate programme—water, world climate programme data and monitoring, WCDMP-45, WMO/TD No. 1013. World Meteorological Organization, Geneva, Switzerland

All India Soil and Land Use Survey (AISLUS) (1990) Watershed atlas of India. All India Soil and Land Use Survey, Ministry of Agriculture, Government of India, New Delhi, India

Andaryani S, Nourani V, Trolle D, Dehgani M, Asl AM (2019) Assessment of land use and climate change effects on land subsidence using a hydrological model and radar technique. J Hydrol 578:124070. https://doi.org/10.1016/j.jhydrol.2019.124070

Andaryani S, Nourani V, Ball J, Jahanbakhsh Asl S, Keshtkar H, Trolle D (2021) A comparison of frameworks for separating the impacts of human activities and climate change on river flow in

S. Mondal and R. K. Paul, *Climate Variability and Human Interference*, Deltas of the World, https://doi.org/10.1007/978-3-032-13257-4

existing records and different near-future scenarios. Hydrol Process 35(7):1–15. https://doi.org/10.1002/hyp.14301

Anil AP, Ramesh H (2017) Analysis of climate trend and effect of land use land cover change on Harangi streamflow, South India: a case study. Sustain Water Resour Manag 3(3):257–267. https://doi.org/10.1007/s40899-017-0088-5

Ardabili S, Mosavi A, Dehghani M, Várkonyi-Kóczy AR (2020) Deep learning and machine learning in hydrological processes climate change and earth systems a systematic review. Lect Notes Netw Syst 101:52–62. https://doi.org/10.1007/978-3-030-36841-8_5

Arnold JG, Moriasi DN, Gassman PW, Abbaspour KC, White MJ, Srinivasan R, Santhi C, Harmel RD, Van Griensven A, Van Liew MW, Kannan N, Jha MK (2012) SWAT: Model use, calibration, and validation. Trans ASABE 55(4):1491–1508

Ashu AB, Lee SIl (2020) Assessing climate change effects on water balance in a monsoon watershed. Water (Switzerland) 12(9). https://doi.org/10.3390/W12092564

Awais M, Vinca A, Parkinson S, McPherson M et al (2022) Replenishing the Indus Delta through multi-sector transformation. Front Environ Sci 10:958101. https://doi.org/10.3389/fenvs.2022.958101

Bai Y, Chen Z, Xie J, Li C (2016) Daily reservoir inflow forecasting using multiscale deep feature learning with hybrid models. J Hydrol 532:193–206. https://doi.org/10.1016/j.jhydrol.2015.11.011

Banko G (1998) A review of assessing the accuracy of classifications of remotely sensed data and of methods including remote sensing data in forest inventory, Interim Report IR-98-081, International Institute for Applied Systems Analysis (IIASA), Laxenburg, Austria

Bao Z, Zhang J, Wang G, Fu G, He R, Yan X, Jin J, Liu Y, Zhang A (2012) Attribution for decreasing streamflow of the Haihe River basin, northern China: climate variability or human activities? J Hydrol 460–461:117–129. https://doi.org/10.1016/j.jhydrol.2012.06.054

Barnett TP, Pierce DW, Hidalgo HG, Bonfils C, Santer BD, Das T, Bala G, Wood AW, Nozawa T, Mirin AA, Cayan DR, Dettinger MD (2008) Human-induced changes in the hydrology of the Western United States. Science 319(5866):1080–1083. https://doi.org/10.1126/SCIENCE.1152538

Bates CG, Henry AJ (1928) Second phase of streamflow experiment at Wagon Wheel Gap, Colo. Mon Weather Rev 56(3)

Beguería S, López-Moreno JI, Lorente A, Seeger M, García-Ruiz JM (2003) Assessing the effect of climate oscillations and land-use changes on streamflow in the central Spanish Pyrenees. Ambio 32(4):283–286. https://doi.org/10.1579/0044-7447-32.4.283

Best J (2019) Anthropogenic stresses on the world's big rivers. Nat Geosci 12(1):7–21. https://doi.org/10.1038/s41561-018-0262-x

Bisai D (2016) Analysing climatological time series of temperature and rainfall for Southern Districts of West Bengal over the period 1901–2011, 1–289

Blum MD, Roberts HH (2009) Drowning of the Mississippi Delta due to insufficient sediment supply and global sea-level rise. Nat Geosci 2:488–491. https://doi.org/10.1038/ngeo553

Bo H, Dong X, Li Z, Hu X, Reta G, Wei C, Su B (2019) Impacts of climate change and human activities on runoff variation of the intensive phosphate mined Huangbaihe River Basin, China. Water (Switzerland) 11(10). https://doi.org/10.3390/w11102039

Bouwer LM, Aerts JCJH, Droogers P, Dolman AJ (2006) Detecting the long-term impacts from climate variability and increasing water consumption on runoff in the Krishna River basin (India). Hydrol Earth Syst Sci 10(5):703–713. https://doi.org/10.5194/HESS-10-703-2006

Brockwell PJ, Davis RA (2016) Introduction to time series and forecasting. https://doi.org/10.1007/978-3-319-29854-2

Budyko MI (1961) On the thermal zones of the Earth (In Russian). Meteorol Hydrol 11

Budyko MI, Miller DH (1974) Climate and life. Academic Press, New York, NY

Buishand TA (1982) Some methods for testing the homogeneity of rainfall records. J Hydrol 58:11–27

Buishand TA (1984) Test for detecting a shift in the mean of hydrological time series. J Hydrol 73:51–69

Buishand TA (1981) The analysis of homogeneity of long-term rainfall records in the Netherland

Bultot F, Dupriez GL, Gellens D (1990) Simulation of land use changes and impacts on the water balance—a case study for Belgium. J Hydrol 114(3–4):327–348. https://doi.org/10.1016/0022-1694(90)90064-5

Census of India (2011) Census of India, Data & Resource, District Census Handbooks

Chang J, Wang Y, Istanbulluoglu E, Bai T, Huang Q, Yang D, Huang S (2015) Impact of climate change and human activities on runoff in the Weihe River Basin, China. Quatern Int 380–381:169–179. https://doi.org/10.1016/j.quaint.2014.03.048

Changnon AS, Demissie M (1996) Detection of changes in streamflow and floods resulting from climate fluctuations and land use-drainage changes definition of precipitation and streamflow changes due to a shift in climate resulting from influences of large cities, global warming, or ev., 411–421

Chaudhary S, Pandey AC (2022) PCA driven watershed prioritization based on runoff modeling and drought severity assessment in parts of Koel River basin, Jharkhand (India). Water Supply 22(2):2034–2054. https://doi.org/10.2166/ws.2021.297

Chemura A, Rwasoka D, Mutanga O, Dube T, Mushore T (2020) The impact of land-use/land cover changes on water balance of the heterogeneous Buzi sub-catchment, Zimbabwe. Remote Sens Appl: Soc Environ 18:100292. https://doi.org/10.1016/J.RSASE.2020.100292

Chen J, Li X, Zhang M (2005) Simulating the impacts of climate variation and land-cover changes on basin hydrology: a case study of the Suomo basin. Sci China, Ser D Earth Sci 48(9):1501–1509. https://doi.org/10.1360/03yd0269

Chen Z, Chen Y, Li B (2013) Quantifying the effects of climate variability and human activities on runoff for Kaidu River Basin in arid region of northwest China. Theoret Appl Climatol 111(3–4):537–545. https://doi.org/10.1007/s00704-012-0680-4

Choudhury P, Rout N, Sahoo P (2007) Impact of sedimentation on environmental flow and compounding natural hazard in the lower catchment area (deltaic environment) of Baitarani River Basin, Orissa

Ciarapica L, Todini E (2002) TOPKAPI: a model for the representation of the rainfall-runoff process at different scales. Hydrol Process 16(2):207–229. https://doi.org/10.1002/hyp.342

Cohen J (1988) The effect size: r. statistical power analysis for the behavioral sciences, 77–83

Cohen J (1992) A power primer. Psychol Bull 112(1):155–159. https://doi.org/10.1037/0033-2909.112.1.155

Cosby BJ, Hornberger GM, Clapp RB, Ginn T (1984) A statistical exploration of the relationships of soil moisture characteristics to the physical properties of soils. Water Resour Res 20(6):682–690

Costa HM, Botta A, Cardille AJ (2003) Effects of large-scale changes in land cover on the discharge of the Tocantins River, Southeastern Amazonia 283:206–217. https://doi.org/10.1016/S0022-1694(03)00267-1

CWC (2014) Brahmani Baitarani basin report. Central Water Commission, Ministry of Water Resources and National Remote Sensing Centre, Govt. of India

D'Agostino RB, Stephens MA (1986) Goodness-of-fit techniques, 560

Darby SE, Hackney CR, Leyland J, Kummu M, Lauri H, Parsons DR, Best JL, Nicholas AP, Aalto R (2016) Fluvial sediment supply to a mega-delta reduced by shifting tropical-cyclone activity. Nature 539(7628):276–279. https://doi.org/10.1038/nature19809

Darby SE, Dunn FE, Nicholls RJ, Rahman M, Riddy L (2015) A first look at the influence of anthropogenic climate change on the future delivery of fluvial sediment to the Ganges–Brahmaputra–Meghna delta. Environ Sci: Processes Impacts, 17:1587–1600. https://doi.org/10.1039/C5EM00252D

Darby SE, Appeaning Addo K, Hazra S, Rahman MM, Nicholls RJ (2020) Fluvial sediment supply and relative sea-level rise. In: Nicholls R, Adger W, Hutton C, Hanson S (eds) Deltas in the Anthropocene. Palgrave Macmillan, Cham, pp 101–127. https://doi.org/10.1007/978-3-030-23517-8_5

Das J, Bhattacharya SK (2018) Trend analysis of long-term climatic parameters in Dinhata of Koch Bihar district, West Bengal. Spat Inf Res 26(3):271–280. https://doi.org/10.1007/S41324-018-0173-3/METRICS

Dey P, Mishra A (2017) Separating the impacts of climate change and human activities on streamflow: a review of methodologies and critical assumptions. J Hydrol 548:278–290. https://doi.org/10.1016/j.jhydrol.2017.03.014

Dijkerman JC (1988) An ustult-aquult-tropept catena in Sierra Leone, West Africa, II. Land qualities and land evaluation. Geoderma 42(1):29–49. https://doi.org/10.1016/0016-7061(88)90021-3

Dong LQ, Zhang GX, Xu YJ (2012) Effects of climate change and human activities on runoff in the Nenjiang River Basin, Northeast China. Hydrol Earth Syst Sci Discuss 9(10):11521–11549. https://doi.org/10.5194/hessd-9-11521-2012

Donohoe A, Marshall J, Ferreira D, Mcgee D (2013) The relationship between ITCZ location and cross-equatorial atmospheric heat transport: from the seasonal cycle to the last glacial maximum. J Clim 26(11):3597–3618. https://doi.org/10.1175/JCLI-D-12-00467.1

Dong W, Cui B, Liu Z, Zhang K (2013) Relative effects of human activities and climate change on the river runoff in an arid basin in northwest China. Hydrol Process 28(18):4854–4864. https://doi.org/10.1002/hyp.9982

Dos Santos V, Laurent F, Abe C, Messner F (2018) Hydrologic response to land use change in a large basin in Eastern Amazon. Water 10(4):429. https://doi.org/10.3390/W10040429

Drummond GB, Tom BDM (2011) Statistics, probability, significance, likelihood: words mean what we define them to mean. J Physiol 589(Pt 16):3901–3904. https://doi.org/10.1113/JPHYSIOL.2011.215103

Du J, Shi C (2012) Effects of climatic factors and human activities on runoff of the Weihe River in recent decades. Quat Int 282:58–65. https://doi.org/10.1016/j.quaint.2012.06.036

Dunford E, Fletcher P (1947) Effect of removal of stream-bank vegetation upon water yield. Trans Am Geophys Union 28(1):105–110

Dwivedi D, Poeppl RE, Wohl E (2025) Hydrological connectivity: a review and emerging strategies for integrating measurement, modeling, and management. Front Water 7:1496199. https://doi.org/10.3389/frwa.2025.1496199

Emery PA (1966) Use of analog model to predict streamflow depletion, Big and Little Blue River Basin, Nebraskaa. Gr Water 4(4):13–19. https://doi.org/10.1111/j.1745-6584.1966.tb01610.x

Engler A (1919) Untersuchungen über den einfluss des waldes auf den stand der gewässer. Mitteilungen Der Schweizerischen Zentralanstalt Fiir Das Forstliche Versuchswesen 12:625

Ercan MB, Maghami I, Bowes BD, Morsy MM, Goodall JL (2020) Estimating potential climate change effects on the Upper Neuse Watershed water balance using the SWAT model. J Am Water Resour Assoc 56(1):53–67. https://doi.org/10.1111/1752-1688.12813

Eschner AR, Satterlund DR (1966) Forest protection and streamflow from an Adirondack Watershed. Water Resour Res 2(4):765–783. https://doi.org/10.1029/WR002i004p00765

Fan J, Tian F, Yang Y, Han S, Qiu G (2010) Quantifying the magnitude of the impact of climate change and human activity on runoff decline in Mian River Basin, China. Water Sci Technol 62(4):783–791. https://doi.org/10.2166/WST.2010.294

Feizabadi S, Li C, Hiatt M (2024) Response of river delta hydrological connectivity to changes in river discharge and atmospheric frontal passage. Front Mar Sci 11:1387180. https://doi.org/10.3389/fmars.2024.1387180

Feng X, Cheng W, Fu B, Lü Y (2016) The role of climatic and anthropogenic stresses on long-term runoff reduction from the Loess Plateau, China. Sci Total Environ 571:688–698. https://doi.org/10.1016/J.SCITOTENV.2016.07.038

Fu G, Charles SP, Chiew FHS (2007) A two-parameter climate elasticity of streamflow index to assess climate change effects on annual streamflow. Water Resour Res 43:W11419. https://doi.org/10.1029/2007WR005890

Gain AK, Rahman MM, Sadik MS, Adnan MSG, Ahmad S, Ahsan SMM, Ashik-Ur-Rahman M, Balke T, Datta DK, Dewan C et al (2022) Overcoming challenges for implementing nature-based solutions in deltaic environments: insights from the Ganges-Brahmaputra delta in Bangladesh. Environ Res Lett 17(6):064052. https://doi.org/10.1088/1748-9326/ac740a

Giambelluca TW (2005) Land use and water resources under a changing climate, 1999

Gibson JJ, Prowse TD, Peters DL (2006) Partitioning impacts of climate and regulation on water level variability in Great Slave Lake. J Hydrol 329(1–2):196–206. https://doi.org/10.1016/J.JHYDROL.2006.02.011

Gogoi K, Rao KN (2022) Analysis of rainfall trends over Assam, North East India. Curr World Environ 17(2):435–446. https://doi.org/10.12944/CWE.17.2.15

Grayson R, Argent R, Nathan R, McMahon T, Mein R (1996) Hydrological recipes estimation techniques in Australian hydrology

GSI (2022) Bhukosh—Geological Survey of India. GSI. https://bhukosh.gsi.gov.in/

Guhathakurta P, Rajeevan M (2008) Trends in the rainfall pattern over India. Int J Climatol 28(11):1453–1469. https://doi.org/10.1002/JOC.1640

Guo J, Liang X, Ruby Leung L (2004) Impacts of different precipitation data sources on water budgets. J Hydrol 298(1–4):311–334. https://doi.org/10.1016/J.JHYDROL.2003.08.020

Guo H, Hu Q, Jiang T (2008) Annual and seasonal streamflow responses to climate and land-cover changes in the Poyang Lake basin, China. J Hydrol 355(1–4):106–122. https://doi.org/10.1016/j.jhydrol.2008.03.020

Guo Y, Li Z, Amo-Boateng M, Deng P, Huang P (2014) Quantitative assessment of the impact of climate variability and human activities on runoff changes for the upper reaches of Weihe River. Stoch Env Res Risk Assess 28(2):333–346. https://doi.org/10.1007/s00477-013-0752-8

Guo S, Zhu Z, Lyu L (2018) Effects of climate change and human activities on soil erosion in the Xihe River Basin, China. Water (Switzerland) 10(8). https://doi.org/10.3390/w10081085

Hajihosseini M, Hajihosseini H, Morid S, Delavar M, Booij MJ (2020) Impacts of land use changes and climate variability on transboundary Hirmand River using SWAT. J Water Clim Change 11(4):1695–1711. https://doi.org/10.2166/wcc.2019.100

Hamed KH, Ramachandra Rao A (1998) A modified Mann-Kendall trend test for autocorrelated data. Jhyd 204(1):182–196. https://doi.org/10.1016/S0022-1694(97)00125-X

Hänsel S, Medeiros DM, Matschullat J, Petta RA, de Mendonça Silva I (2016) Assessing homogeneity and climate variability of temperature and precipitation series in the capitals of North-Eastern Brazil. Front Earth Sci 4(March):1–21. https://doi.org/10.3389/feart.2016.00029

Hawkins DM (1977) Testing a sequence of observations for a shift in location. J Am Stat Assoc 72(357):180–186. https://doi.org/10.1080/01621459.1977.10479935

Helsel DR, Hirsch RM, Ryberg KR, Archfield SA, Gilroy EJ (2020). Statistical methods in water resources. Techn Methods 2020(4-A3):1–484. https://doi.org/10.3133/TM4A3

Helsel DR, Hirsch RM (1995) Statistical methods in water resources. Elsevier, Amsterdam

Henning D, Flohn F (1977) Climate aridity index map, United Nations Conferenceon Desertification, FAO and WMO

Heo J, Yu J, Giardino JR, Cho H (2015a) Impacts of climate and land-cover changes on water resources in a humid subtropical watershed: a case study from East Texas, USA. Water Environ J 29(1):51–60. https://doi.org/10.1111/wej.12096

Heo J, Yu J, Giardino JR, Cho H (2015b) Water resources response to climate and land-cover changes in a semi-arid watershed, New Mexico, USA 26(4):463–474. https://doi.org/10.3319/TAO.2015.03.24.01(Hy)1

Hiatt M, Passalacqua P (2015) Hydrological connectivity in river deltas: the first-order importance of channel-island exchange. Water Resour Res 51:2264–2282. https://doi.org/10.1002/2014WR016149

Hibbert AR (1967) Forest treatment effects on water yield. In: International symposium for hydrology, pp 527–543

Hipel KW, McLeod AI (1994a) Time series modelling of water resources and environmental systems, development in water science 45. Elsevier

Hipel KW, McLeod AI (1994b) Time series modelling of water resources and environmental systems, 1013

Hirsch RM, Slack JR, Smith RA (1982) Techniques of trend analysis for monthly water quality data. Water Resour Res 18(1):107–121. https://doi.org/10.1029/WR018I001P00107

Hirsch RM, Slack JR, Smith RA (1982) Techniques of trend analysis for monthly water-quality data. Water Resour Res 18(1):107–121

Hoover MD (1944) Effect of removal forest vegetation upon water-yields. Eos, Trans Am Geophys Usnion 25(6):969–977. https://doi.org/10.1029/TR025I006P00969

Hoyt W, Troxell H (1934) Forests and stream flow. ASCE 99(Paper No. 1858):1–111

Hu S, Liu C, Zheng H, Wang Z, Yu J (2012) Assessing the impacts of climate variability and human activities on streamflow in the water source area of Baiyangdian Lake. J Geog Sci 22(5):895–905. https://doi.org/10.1007/S11442-012-0971-9

Hu J, Ma J, Nie C, Xue L, Zhang Y, Ni F, Deng Y, Liu J, Zhou D, Li L, Wang Z (2020) Attribution analysis of runoff change in Min-Tuo River Basin based on SWAT model simulations, China. Sci Rep 10(1):1–16. https://doi.org/10.1038/s41598-020-59659-z

Huo Z, Feng S, Kang S, Li W, Chen S (2008) Effect of climate changes and water-related human activities on annual stream flows of the Shiyang river basin in arid north-west China. Hydrol Process 22(16):3155–3167. https://doi.org/10.1002/HYP.6900

IMD (2020) Observed monsoon rainfall variability and changes during recent 30 years (1989–2018). Government of India, Ministry of Earth Sciences (MoES), Climate Research and Services (CRS) Division, Pune, 1–10

India-WRIS (2020) Water data, India Water Resources Information System (India-WRIS). Department of Water Resources, RD & GR, Ministry of Jal Shakti, Govt. of India. https://indiawris.gov.in/wris/#/

ISRIC-WDC (2020) International Soil Reference and Information Centre (ISRIC), World Data System (WDS), Wageningen University & Research in Wageningen, The Netherlands

Jiang H, Chen J, Guan Q, Dong L, Fu C (2011b) Comparison of monthly runoff for different scaling basin and sensitivity analysis: a case study in the Xijiang Watershed [In Chinese]. J Subtrop Res Environ 6(1):45–55

Jiang S, Ren L, Yong B, Singh VP, Yang X, Yuan F (2011a) Quantifying the effects of climate variability and human activities on runoff from the Laohahe basin in northern China using three different methods. Hydrol Process 25(16):2492–2505. https://doi.org/10.1002/HYP.8002

Jiang C, Xiong L, Wang D, Liu P, Guo S, Xu CY (2015) Separating the impacts of climate change and human activities on runoff using the Budyko-type equations with time-varying parameters. J Hydrol 522:326–338. https://doi.org/10.1016/j.jhydrol.2014.12.060

Jiao Y, Liu J, Li C, Wang W, Yu F, Wang Y (2020) Quantitative attribution of runoff attenuation to climate change and human activity in typical mountainous areas: an enlightenment to water resource sustainable utilization and management in North China. Sustainability (Switzerland) 12(24):1–19. https://doi.org/10.3390/su122410395

Jones RN, Chiew FHS, Boughton WC, Zhang L (2006) Estimating the sensitivity of mean annual runoff to climate change using selected hydrological models. Adv Water Resour 29(10):1419–1429. https://doi.org/10.1016/j.advwatres.2005.11.001

Juckem PF, Hunt RJ, Anderson MP, Robertson DM (2008) Effects of climate and land management change on streamflow in the driftless area of Wisconsin. J Hydrol 355(1–4):123–130. https://doi.org/10.1016/J.JHYDROL.2008.03.010

Kalcic M, Chaubey I, Frankenberger J (2015) Defining Soil and Water Assessment Tool (SWAT) hydrologic response units (HRUs) by field boundaries. Ijabe. Org 8(3):69–80. https://doi.org/10.3965/j.ijabe.20150803.951

Kang Y, Gao J, Shao H, Zhang Y (2020) Quantitative analysis of hydrological responses to climate variability and land-use change in the hilly-gully region of the loess plateau, China. Water (Switzerland) 12(1). https://doi.org/10.3390/w12010082

Kang H, Sridhar V, Ali SA (2022) Climate change impacts on conventional and flash droughts in the Mekong River Basin. Sci Total Environ 838(Part 2):155845. https://doi.org/10.1016/j.scitotenv.2022.155845

Karl TR, Riebsame WE (1989) The impact of decadal fluctuations in mean precipitation and temperature on runoff: a sensitivity study over the United States. Clim Change 15(3):423–447. https://doi.org/10.1007/BF00240466

Karunanithi BN, Grenney WJ, Whitley D, Bovee K (1994) (1966) Parametric models are expensive and time consuming to apply, and are normally employed only when insufficient data exist to utilize a simpler empirical method. Sufficient data were available in this situation to apply the most commonly used empi. 8(2):201–220

Kendall MG (1975) Rank correlation methods, 4th edn. Charles Griffin

Kendall MG, Gibbons JD (1990) Rank correlation methods, 5th edn. Edward Arnold

Kendall MG (1975) Rank correlation methods, 4th editio. Charles Griffin, London

Kezer K, Matsuyama H (2006) Decrease of river runoff in the Lake Balkhash basin in Central Asia. Hydrol Process 20(6):1407–1423. https://doi.org/10.1002/hyp.6097

Khaliq MN, Ouarda TBMJ (2007) On the critical values of the standard normal homogeneity test (SNHT). Int J Climatol 27(5):681–687. https://doi.org/10.1002/JOC.1438

Khoi DN, Nguyen VT, Sam TT, Nhi PTT (2019) Evaluation on effects of climate and land-use changes on streamflow and water quality in the La Buong River Basin, Southern Vietnam. Sustainability (Switzerland) 11(24). https://doi.org/10.3390/SU11247221

Kondolf GM et al (2014a) Sustainable sediment management in reservoirs and regulated rivers: experiences from five continents. Earth's Future 2(5):256–280. https://doi.org/10.1002/2013EF000184

Kondolf GM, Rubin ZK, Minear JT (2014b) Dams on the Mekong: cumulative sediment starvation. Water Resour Res 50(6):5158–5169. https://doi.org/10.1002/2013WR014651

Kratzert F, Klotz D, Brenner C, Schulz K, Herrnegger M (2018) Rainfall-runoff modelling using Long Short-Term Memory (LSTM) networks. Hydrol Earth Syst Sci 22(11):6005–6022. https://doi.org/10.5194/HESS-22-6005-2018

Kristian J, Christian J, Mazvimavi D (1998) Assessing the effect of land use change on catchment runoff by combined use of statistical tests and hydrological modelling: case studies from Zimbabwe 205:147–163

Kulkarni A, Sabin TP, Chowdary JS, KoteswaraRao K, Priya P, Gandhi N, Bhaskar P, Buri VK, Sabade SS (2020) Precipitation changes in India. Assessment of Climate Change over the Indian

Region: A Report of the Ministry of Earth Sciences (MoES), Government of India, 47–72. https://doi.org/10.1007/978-981-15-4327-2_3/FIGURES/4

Kumar V, Jain SK, Singh Y (2010) Analysis of long-term rainfall trends in India. Hydrol Sci J 55(4):484–496. https://doi.org/10.1080/02626667.2010.481373

Kundzewicz ZW, Robson AJ (2004) Change detection in hydrological records—a review of the methodology. Hydrol Sci J 49(1):7–19. https://doi.org/10.1623/hysj.49.1.7.53993

Langbein WB, Wrather WE (1949) Annual runoff in the United States (Issue June). https://doi.org/10.3133/cir52

Lee S, Kim SU (2017) Quantification of hydrological responses due to climate change and human activities over various time scales in South Korea. Water (Switzerland) 9(1). https://doi.org/10.3390/w9010034

Legesse D, Vallet-Coulomb C, Gasse F (2003) Hydrological response of a catchment to climate and land use changes in Tropical Africa: case study South Central Ethiopia. J Hydrol 275(1–2):67–85. https://doi.org/10.1016/S0022-1694(03)00019-2

Li LJ, Zhang L, Wang H, Wang J, Yang JW, Jiang DJ, Li JY, Qin DY (2007a) Assessing the impact of climate variability and human activities on streamflow from the Wuding River basin in China. Hydrol Process 21(25):3485–3491. https://doi.org/10.1002/HYP.6485

Li L-J, Zhang L, Wang H, Wan J, Yang J-W, Jiang D-J, Li J-Y, Qin D-Y (2007b) Assessing the impact of climate variability and human activities on streamflow from the Wuding River basin in China. Hydrol Process 21:3485–3491. https://doi.org/10.1002/hyp.6485

Li LJ, Li B, Liang LG, Li JY, Liu YM (2010) Effect of climate change and land use on stream flow in the upper and middle reaches of the Taoer River, northeastern China. For Stud China 12(3):107–115. https://doi.org/10.1007/s11632-010-0301-1

Li H, Zhang Y, Vaze J, Wang B (2012) Separating effects of vegetation change and climate variability using hydrological modelling and sensitivity-based approaches. J Hydrol 420–421:403–418. https://doi.org/10.1016/j.jhydrol.2011.12.033

Li B, Su H, Chen F, Li H, Zhang R, Tian J, Chen S, Yang Y, Rong Y (2013) Separation of the impact of climate change and human activity on streamflow in the upper and middle reaches of the Taoer River, northeastern China. Theoret Appl Climatol 118(1–2):271–283. https://doi.org/10.1007/s00704-013-1032-8

Li J, Li G, Zhou S, Chen F (2015) Quantifying the effects of land surface change on annual runoff considering precipitation variability by SWAT. Water Resour Manage 30(3):1071–1084. https://doi.org/10.1007/s11269-015-1211-8

Li S, Yang G, Wang H (2019a) The runoff evolution and the differences analysis of the causes of runoff change in different regions: a case of the Weihe River Basin, Northern China. Sustainability (Switzerland) 11(19):1–24. https://doi.org/10.3390/su11195295

Li Y, Chang J, Luo L, Wang Y, Guo A, Ma F, Fan J (2019b) Spatiotemporal impacts of land use land cover changes on hydrology from the mechanism perspective using SWAT model with time-varying parameters. Hydrol Res 50(1):244–261. https://doi.org/10.2166/nh.2018.006

Li B, Shi X, Lian L, Chen Y, Chen Z, Sun X (2020a) Quantifying the effects of climate variability, direct and indirect land use change, and human activities on runoff. J Hydrol 584:124684. https://doi.org/10.1016/j.jhydrol.2020.124684

Li H, Shi C, Zhang Y, Ning T, Sun P, Liu X, Ma X, Liu W, Collins AL (2020b) Using the Budyko hypothesis for detecting and attributing changes in runoff to climate and vegetation change in the soft sandstone area of the middle Yellow River basin, China. Sci Total Environ 703:135588. https://doi.org/10.1016/J.SCITOTENV.2019.135588

Li D, Gao W, Shao D, Amenuvor M, Tong Y, Cui B (2021) A tale of two deltas: dam-induced hydromorphological evolution of the Volta River Delta (Ghana) and Yellow River Delta (China). Water 13(22):3198. https://doi.org/10.3390/w13223198

Li Z, Liu W, Zhang X, Zheng F (2009) Impacts of land use change and climate variability on hydrology in an agricultural catchment on the Loess Plateau of China. J Hydrol 377(1–2):35–42. https://doi.org/10.1016/J.JHYDROL.2009.08.007

Li Y, Chang J, Wang Y, Jin W, Guo A (2016) Spatiotemporal impacts of climate, land cover change and direct human activities on runoff variations in the Wei River Basin, China. Water (Switzerland) 8(6). https://doi.org/10.3390/w8060220

Liang W, Bai D, Wang F, Fu B, Yan J, Wang S, Yang Y, Long D, Feng M (2015) Quantifying the impacts of climate change and ecological restoration on streamflow changes based on a Budyko hydrological model in China's Loess Plateau. Water Resour Res 51(8):6500–6519. https://doi.org/10.1002/2014WR016589

Liang K, Liu C, Liu X (2013) Impacts of climate variability and human activity on streamflow decrease in a sediment concentrated region in the Middle Yellow River, 1741–1749. https://doi.org/10.1007/s00477-013-0713-2

Liu Q, Yang Z, Cui B, Sun T (2009) Temporal trends of hydro-climatic variables and runoff response to climatic variability and vegetation changes in the Yiluo River basin, China. Hydrol Process 23(21):3030–3039. https://doi.org/10.1002/HYP.7414

Liu D, Chen X, Lian Y, Lou Z (2010) Impacts of climate change and human activities on surface runoff in the Dongjiang River basin of China. Hydrol Process 24(11):1487–1495. https://doi.org/10.1002/hyp.7609

Liu Y, Zhang X, Xia D, You J, Rong Y, Bakir M (2013) Impacts of land-use and climate changes on hydrologic processes in the Qingyi River Watershed, China. J Hydrol Eng 18(11):1495–1512. https://doi.org/10.1061/(ASCE)HE.1943-5584.0000485

Liu J, Zhou Z, Yan Z, Gong J, Jia Y, Xu CY, Wang H (2019) A new approach to separating the impacts of climate change and multiple human activities on water cycle processes based on a distributed hydrological model. J Hydrol 578(August):124096. https://doi.org/10.1016/j.jhydrol.2019.124096

López-Moreno JI, Zabalza J, Vicente-Serrano SM, Revuelto J, Gilaberte M, Azorin-Molina C, Morán-Tejeda E, García-Ruiz JM, Tague C (2014) Impact of climate and land use change on water availability and reservoir management: scenarios in the Upper Aragón River, Spanish Pyrenees. Sci Total Environ 493:1222–1231. https://doi.org/10.1016/j.scitotenv.2013.09.031. Epub 2013 Oct 3. PMID: 24090497

Lorrai M, Sechi GM (1995) Neural nets for modelling rainfall-runoff transformations. Water Resour Manage 9(4):299–313. https://doi.org/10.1007/BF00872489

Lørup JK, Refsgaard JC, Mazvimavi D (1998) Assessing the effect of land use change on catchment runoff by combined use of statistical tests and hydrological modelling: case studies from Zimbabwe. J Hydrol 205(3–4):147–163. https://doi.org/10.1016/S0168-1176(97)00311-9

Luo L, Zhou Q, He HS, Duan L, Zhang G, Xie H(2020) Relative importance of land use and climate change on hydrology in agricultural watershed of southern China. Sustainability (Switzerland) 12(16). https://doi.org/10.3390/SU12166423

Lyu L, Wang X, Sun C, Ren T, Zheng D (2019) Quantifying the effect of land use change and climate variability on greenwater resources in the Xihe River Basin, Northeast China. Sustainability (Switzerland) 11(2). https://doi.org/10.3390/su11020338

Ma Z, Kang S, Zhang L, Tong L, Su X (2008) Analysis of impacts of climate variability and human activity on streamflow for a river basin in arid region of northwest China. J Hydrol 352(3–4):239–249. https://doi.org/10.1016/j.jhydrol.2007.12.022

Ma X, Xu J, Luo Y, Aggarwal SP, Li J (2009) Response of hydrological processes to land-cover and climate changes in Kejie watershed, south-west China. Hydrol Process 23(8):1179–1191. https://doi.org/10.1002/HYP.7233

Ma H, Yang D, Tan SK, Gao B, Hu Q (2010) Impact of climate variability and human activity on streamflow decrease in the Miyun Reservoir catchment. J Hydrol 389(3–4):317–324. https://doi.org/10.1016/J.JHYDROL.2010.06.010

Ma K, Huang X, Liang C, Zhao H, Zhou X, Wei X (2020b) Effect of land use/cover changes on runoff in the Min River watershed. River Res Appl 36(5):749–759. https://doi.org/10.1002/rra.3608

Ma D, Xu YP, Xuan W, Gu H, Sun Z, Bai Z (2020) Do model parameters change under changing climate and land use in the upstream of the Lancang River Basin, China? Hydrol Sci J, 1894–1908. https://doi.org/10.1080/02626667.2020.1782915

Mahmoodi N, Wagner PD, Kiesel J, Fohrer N (2021) Modeling the impact of climate change on streamflow and major hydrological components of an Iranian Wadi system. J Water Clim Change 12(5):1598–1613. https://doi.org/10.2166/wcc.2020.098

Mann HB (1945) Non-parametric test against trend. Econometrica 13(3):245–259

McCabe GJ, Wolock DM (2002) A step increase in streamflow in the conterminous United States. Geophys Res Lett 29(24):38–41. https://doi.org/10.1029/2002GL015999

McCuen RH, Knight Z, Cutter AG (2006) Evaluation of the Nash-Sutcliffe efficiency index. J Hydrol Eng 11(6):597–602. https://doi.org/10.1061/(ASCE)1084-0699(2006)11:6(597)

Meng F, Liu T, Huang Y, Luo M, Bao A, Hou D (2016) Quantitative detection and attribution of runoff variations in the Aksu River Basin. Water (Switzerland) 8(8). https://doi.org/10.3390/w8080338

Meybeck M, Vörösmarty C (2005) Fluvial filtering of land-to-ocean fluxes: from natural Holocene variations to the Anthropocene. CR Geosci 337(1–2):107–123. https://doi.org/10.1016/j.crte.2004.09.016

Milly PCD (1994) Climate, interseasonal storage of soil water, and the annual water balance. Adv Water Resour 17(1–2):19–24. https://doi.org/10.1016/0309-1708(94)90020-5

Milly PCD, Dunne KA (2002) Macroscale water fluxes 2. Water and energy supply control of their interannual variability. Water Resour Res 38(10):1206. https://doi.org/10.1029/2001WR000760

Mishra P, Pandey CM, Singh U, Gupta A, Sahu C, Keshri A (2019) Descriptive statistics and normality tests for statistical data. Ann Card Anaesth 22(1):67. https://doi.org/10.4103/ACA.ACA_157_18

Mishra SP, Kumar C, Mishra A, Mishra S, Patel A (2021) Estimation of sediments in Rengali Reservoir, Odisha (India) using remote sensing. Int J Environ Clim Change 11(12):205–225. https://doi.org/10.9734/ijecc/2021/v11i1230572

Molle F (2009) River-basin planning and management: the social life of a concept. Geoforum 40(3):484–494. https://doi.org/10.1016/j.geoforum.2009.03.004

Montenegro A & Ragab R (2010) Hydrological response of a Brazilian semi-arid catchment to different land use and climate change scenarios: a modelling study. Hydrol Process 24(19):2705–2723. https://doi.org/10.1002/hyp.7825

Morán-Tejeda E, Ceballos-Barbancho A, Llorente-Pinto JM (2010) Hydrological response of Mediterranean headwaters to climate oscillations and land-cover changes: the mountains of Duero River basin (Central Spain). Global Planet Change 72(1–2):39–49. https://doi.org/10.1016/J.GLOPLACHA.2010.03.003

Moshe Z, Metzger A, Elidan G, Kratzert F, Nevo S, El-Yaniv R (2020) HydroNets: leveraging River structure for hydrologic modeling. arXiv preprint. https://arxiv.org/abs/2007.00595

Musie M, Sen S, Chaubey I (2020) Hydrologic responses to climate variability and human activities in Lake Ziway Basin, Ethiopia. Water (Switzerland) 12(1):1–26. https://doi.org/10.3390/w12010164

Napoli M, Massetti L, Orlandini S (2017) Hydrological response to land use and climate changes in a rural hilly basin in Italy. Catena 157(2016):1–11. https://doi.org/10.1016/j.catena.2017.05.002

NASA (2020) Weather data, National Aeronautics and Space Administration (NASA) Langley Research Center (LaRC) Prediction of Worldwide Energy Resource (POWER) Project funded through the NASA Earth Science/Applied Science Program. https://power.larc.nasa.gov/data-acce

NASA Power data access (2020) Weather data, National Aeronautics and Space Administration (NASA) Langley Research Center (LaRC) Prediction of Worldwide Energy Resource (POWER) Project funded through the NASA Earth Science/Applied Science Program. https://power.larc.nasa.gov/data-acce

Nash LL, Gleick PH (1991) Sensitivity of streamflow in the Colorado Basin to climatic changes. J Hydrol 125(3–4):221–241. https://doi.org/10.1016/0022-1694(91)90030-L

Nash JE, Sutcliffe JV (1970) River flow forecasting through conceptual models part I—a discussion of principles. J Hydrol 10(3):282–290. https://doi.org/10.1016/0022-1694(70)90255-6

Natkhin M, Dietrich O, Schäfer MP, Lischeid G (2015) The effects of climate and changing land use on the discharge regime of a small catchment in Tanzania. Reg Environ Change 15(7):1269–1280. https://doi.org/10.1007/s10113-013-0462-2

Neitsch SL, Arnold JG, Kiniry JR, Williams JR (2011) Soil & water assessment tool theoretical documentation version 2009: Texas Water Resources Institute Technical Report No. 406

Nepal S, Flügel WA, Shrestha AB (2014) Upstream-downstream linkages of hydrological processes in the Himalayan region. Ecol Process 3:19. https://doi.org/10.1186/s13717-014-0019-4

Nguyen HTT, Doan TM, Tomppo E, McRoberts RE (2020) Land Use/land cover mapping using multitemporal Sentinel-2 imagery and four classification methods—a case study from Dak Nong, Vietnam. Remote Sens 12(9):1367. https://doi.org/10.3390/RS12091367

Nie N, Zhang W, Liu M, Chen H, Zhao D (2021) Separating the impacts of climate variability, land-use change and large reservoir operations on streamflow in the Yangtze River basin, China, using a hydrological modeling approach. Int J Digit Earth 14(2):231–249. https://doi.org/10.1080/17538947.2020.1812740

Ning T, Li Z, Liu W (2016) Separating the impacts of climate change and land surface alteration on runoff reduction in the Jing River catchment of China. CATENA 147:80–86. https://doi.org/10.1016/j.catena.2016.06.041

Nkhoma L, Ngongondo C, Dulanya Z, Monjerezi M (2021) Evaluation of integrated impacts of climate and land use change on the river flow regime in wamkurumadzi river, shire basin in malawi. J Water Clim Change 12(5):1674–1693. https://doi.org/10.2166/wcc.2020.138

Nourani V, Roughani A, Gebremichael M (2011) TOPMODEL capability for rainfall-runoff modeling of the Ammameh watershed at different time scales using different terrain algorithms. J Urban Environ Eng 5(1):1–14. http://www.jstor.org/stable/26203351

Omer A, Wang W, Basheer AK, Yong B (2016) Integrated assessment of the impacts of climate variability and anthropogenic activities on river runoff: a case study in the Hutuo River Basin, China. Hydrol Res 48(2):416–430. https://doi.org/10.2166/nh.2016.229

Onyutha C, Nyesigire R, Nakagiri A (2021) Contributions of human activities and climatic variability to changes in river rwizi flows in uganda, east africa. Hydrology 8(4). https://doi.org/10.3390/hydrology8040145

Palmer MA, Reidy Liermann CA, Nilsson C, Flörke M, Alcamo J, Lake PS, Bond N (2008) Climate change and the world's river basins: anticipating management options. Front Ecol Environ 6(2):81–89. https://doi.org/10.1890/060148

Pandey AC (2017) Geospatial technique for runoff estimation based on SCS-CN method in Upper South Koel River Basin of Jharkhand (India). Int J Hydrol 1(7). https://doi.org/10.15406/ijh.2017.01.00037

Patakamuri SK, O'Brien N (2020) Modifiedmk: modified Mann Kendall trend tests. 10.1023/B

Patakamuri SK, Muthiah K, Sridhar V (2020) Long-term homogeneity, trend, and change-point analysis of rainfall in the arid district of ananthapuramu, Andhra Pradesh State, India. Water (Switzerland) 12(1). https://doi.org/10.3390/w12010211

Patterson LA, Lutz B, Doyle MW (2013) Climate and direct human contributions to changes in mean annual streamflow in the South Atlantic. USA 49(August):7278–7291. https://doi.org/10.1002/2013WR014618

Penman HL (1948) Natural evaporation from open water, bare soil and grass. Proc R Soc Lond Ser A Math Phys Sci or Proc R Soc A 193(1032):120–145. https://doi.org/10.1098/rspa.1948.0037

Peña-Arancibia JL, van Dijk AIJM, Guerschman JP, Mulligan M, (Sampurno) Bruijnzeel LA, McVicar TR (2012) Detecting changes in streamflow after partial woodland clearing in two large catchments in the seasonal tropics. J Hydrol, 416–417:60–71. https://doi.org/10.1016/J.JHYDROL.2011.11.036

Peng S, Liu W, Wang W, Shao Q, Jiao X, Yu Z, Xing W, Xu J, Zhang Z, Luo Y (2013) Estimating the effects of climatic variability and human activities on streamflow in the Hutuo River Basin, China. J Hydrol Eng 18(4):422–430. https://doi.org/10.1061/(asce)he.1943-5584.0000664

Pérez S, Sierra E, López E, Nizzero G, Momo F, Massobrio M (2011) Abrupt changes in rainfall in the Eastern area of La Pampa Province, Argentina. Theor Appl Climatol 103(1–2):159–165. https://doi.org/10.1007/s00704-010-0290-y

Petchprayoon P, Blanken PD, Ekkawatpanit C, Hussein K (2010) Hydrological impacts of land use/land cover change in a large river basin in central-northern Thailand. Int J Climatol 30(13):1917–1930. https://doi.org/10.1002/joc.2131

Pettitt AN (1979) A non-parametric to the approach problem. Appl Stat 28(2):126–135

Petz B (1981) Basic statistical methods for non-mathematicians. Liber, Zagreb, 409 (in Croatian)

Piliouras A, Kim W (2019) Upstream and downstream boundary conditions control the physical and biological development of river deltas. Geophys Res Lett 46(20):11385–11393. https://doi.org/10.1029/2019GL084045

Pirnia A, Darabi H, Choubin B, Omidvar E, Onyutha C, Haghighi AT (2019) Contribution of climatic variability and human activities to stream flow changes in the Haraz River basin, northern Iran. J Hydro-Environ Res, 25(2017):12–24. https://doi.org/10.1016/j.jher.2019.05.001

Poelmans L, Rompaey AV, Ntegeka V, Willems P (2011) The relative impact of climate change and urban expansion on peak flows: a case study in central Belgium. Hydrol Process 25(18):2846–2858. https://doi.org/10.1002/HYP.8047

Poff NL, Schmidt JC (2016) How dams can go with the flow. Science 353(6304):1099–1100. https://doi.org/10.1126/science.aah4926

Potter NJ, Zhang L, Milly PCD, McMahon TA, Jakeman AJ (2005) Effects of rainfall seasonality and soil moisture capacity on mean annual water balance for Australian catchments. Water Resour Res 41:W06007. https://doi.org/10.1029/2004WR003697

Praveen B, Talukdar S, Shahfahad, Mahato S, Mondal J, Sharma P, Islam ARMT, Rahman A (2020) Analyzing trend and forecasting of rainfall changes in India using non-parametrical and machine learning approaches. Sci Rep 10(1):1–21. https://doi.org/10.1038/s41598-020-67228-7

Raff JL, Goodbred SL Jr, Pickering JL et al (2023) Sediment delivery to sustain the Ganges-Brahmaputra delta under climate change and anthropogenic impacts. Nat Commun 14:2429. https://doi.org/10.1038/s41467-023-38057-9

Raha GN, Bhattacharjee K, Das M, Dutta M, Bandyopadhyay S (2014) Statistical study of surface temperature and rainfall over four stations in north Bengal. MAUSAM, 65(2):179–184. https://doi.org/10.54302/MAUSAM.V65I2.969

Rahman M et al (2018) Recent sediment flux to the Ganges-Brahmaputra-Meghna delta system. Sci Total Environ 643:1054–1064. https://doi.org/10.1016/j.scitotenv.2018.06.258

Refsgaard JC, Alley WM, Vuglinsky VS (1989) Methodology for distinguishing between man's influence and climatic effects on the hydrological cycle. SC.89/WS/22

Rivas-Tabares D, Tarquis AM, De Miguel Á, Gobin A, Willaarts B (2022) Enhancing LULC scenarios impact assessment in hydrological dynamics using participatory mapping protocols in semiarid regions. Sci Total Environ 803:149906. https://doi.org/10.1016/j.scitotenv.2021.149906

Sahin V, Hall MJ (1996) The effects of afforestation and deforestation on water yields. J Hydrol 178(1–4):293–309. https://doi.org/10.1016/0022-1694(95)02825-0

Saifullah M, Li Z, Li Q, Zaman M, Hashim S (2016) Quantitative estimation of the impact of precipitation and land surface change on hydrological processes through statistical modeling. Adv Meteorol 2016. https://doi.org/10.1155/2016/6130179

Sao D, Kato T, Tu LH, Thouk P, Fitriyah A, Oeurng C (2020) Evaluation of different objective functions used in the SUFI-2 calibration process of SWAT-CUP on water balance analysis: a case study of the Pursat River Basin, Cambodia. Water 12(10):2901. https://doi.org/10.3390/W12102901

Sawilowsky SS (2009) New effect size rules of thumb. J Mod Appl Stat Methods 8(2):26. https://doi.org/10.22237/jmasm/1257035100

Schaake JC (1990) From climate to flow. In: Waggoner PE (ed) Climate change and U.S. water resources, chap 8. John Wiley, New York, pp 177–206

Schmitt RJP, Bizzi S, Castelletti A, Opperman JJ, Kondolf GM (2019) Planning dam portfolios for low sediment trapping shows limits for sustainable hydropower in the Mekong. Sci Adv 5(10):eaaw2175. https://doi.org/10.1126/sciadv.aaw2175

Schmitt RJP, Giuliani M, Bizzi S, Kondolf GM, Daily GC, Castelletti A (2021) Strategic basin and delta planning increases the resilience of the Mekong Delta under future uncertainty. Proc Natl Acad Sci 118(36):e2026127118. https://doi.org/10.1073/pnas.2026127118

Sen PK (1968) Estimates of the regression coefficient based on Kendall's Tau. J Am Stat Assoc 63(324):1379–1389. https://doi.org/10.1080/01621459.1968.10480934

Setyorini A, Khare D, Pingale SM (2017) Simulating the impact of land use/land cover change and climate variability on watershed hydrology in the Upper Brantas basin, Indonesia. Appl Geomatics 9(3):191–204. https://doi.org/10.1007/s12518-017-0193-z

Seyoum WM, Milewski AM, Durham MC (2015) Understanding the relative impacts of natural processes and human activities on the hydrology of the Central Rift Valley lakes, East Africa. Hydrol Process 29(19):4312–4324. https://doi.org/10.1002/hyp.10490

Shang X, Jiang X, Jia R, Wei C (2019) Land use and climate change effects on surface runoff variations in the upper Heihe River basin. Water (Switzerland) 11(2). https://doi.org/10.3390/w11020344

Shapiro SS, Wilk MB (1965) An analysis of variance test for normality (complete samples). Biometrika 52(3/4):591. https://doi.org/10.2307/2333709

Sharma PJ, Patel PL, Jothiprakash V (2019) Impact of rainfall variability and anthropogenic activities on streamflow changes and water stress conditions across Tapi Basin in India. Sci Total Environ 687:885–897. https://doi.org/10.1016/J.SCITOTENV.2019.06.097

Sharma S, Singh PK (2017) Long term spatiotemporal variability in rainfall trends over the State of Jharkhand, India. Climate 5(1):18. https://doi.org/10.3390/CLI5010018

Shen C (2018) A transdisciplinary review of deep learning research and its relevance for water resources scientists. Water Resour Res 54(11):8558–8593. https://doi.org/10.1029/2018WR022643

Shi P, Hou Y, Xie Y, Chen C, Chen X, Li Q, Qu S, Fang X, Srinivasan R (2013a) Application of a SWAT model for hydrological modeling in the Xixian Watershed, China. J Hydrol Eng 18(11):1522–1529. https://doi.org/10.1061/(asce)he.1943-5584.0000578

Shi P, Ma X, Hou Y, Li Q, Zhang Z, Qu S, Chen C, Cai T, Fang X (2013b) Effects of land-use and climate change on hydrological processes in the Upstream of Huai River, China. Water Resour Manag 27(5):1263–1278. https://doi.org/10.1007/s11269-012-0237-4

Shuttle Radar Topography Mission Digital Elevation Model (SRTM DEM) (2022) SRTM Global 1 arc-second [Digital elevation model]. U.S. Geological Survey (USGS), EarthExplorer. https://earthexplorer.usgs.gov/

Singh VP (1994) Elementary hydrology. Prentice Hall of India Private Limited. Retrieved from http://ecoursesonline.iasri.res.in/mod/page/view.php?id=125271

Sinha PN (2017) An alternative approach to evaluating inter-basin water transfer links: a case study of the Inter-Linking of Rivers Project in India. 348

Sneyers R (1990) On the statistical analysis of series of observations. World Meteorological Organization, Technical Note No. 143, Geneva

Soil Survey Division Staff (1993) Soil survey manual. Soil Conservation Service, U.S. Department of Agriculture Handbook 18

Stanley DJ, Warne AG (1998) Nile Delta in its destruction phase. J Coastal Res 14(3):794–825

Stuti, Pandey AC, Gupta SK (2020) Groundwater prospect mapping in Upper South Koel River Basin Jharkhand (India) based on Gis and Remote Sensing Technologies. In: Wastewater reuse and watershed management, May, 459–480. https://doi.org/10.1201/9780429433986-38

Sankarasubramanian A, Vogel RM, Limbrunner JF (2001) Climate elasticity of streamflow in the United States. Water Resour Res 37(6):1771–1781. https://doi.org/10.1029/2000WR900330

Sun Y, Tian F, Yang L, Hu H (2014) Exploring the spatial variability of contributions from climate variation and change in catchment properties to streamflow decrease in a mesoscale basin by three different methods. J Hydrol 508:170–180. https://doi.org/10.1016/j.jhydrol.2013.11.004

Sun P, Wu Y, Wei X, Sivakumar B, Qiu L, Mu X, Chen J, Gao J (2020) Quantifying the contributions of climate variation, land use change, and engineering measures for dramatic reduction in streamflow and sediment in a typical loess watershed, China. Ecol Eng 142(May 2019):105611. https://doi.org/10.1016/j.ecoleng.2019.105611

Swain JB, Patra KC (2019) Impact assessment of land use/land cover and climate change on streamflow regionalization in an ungauged catchment. J Water Clim Change 10(3):554–568. https://doi.org/10.2166/wcc.2018.161

Syvitski JPM (2010) Impacts of sediment retention by dams on delta shoreline recession: evidences from the Krishna and Godavari deltas, India. Earth Surf Process Land 35(7):817–827. https://doi.org/10.1002/esp.1977

Syvitski JP et al (2009) Sinking deltas due to human activities. Nat Geosci 2(10):681–686. https://doi.org/10.1038/ngeo629

Szcześniak M, Piniewski M (2015) Improvement of hydrological simulations by applying daily precipitation interpolation schemes in meso-scale catchments. Water 7(2):747–779. https://doi.org/10.3390/W7020747

Tang Y, Tang Q, Tian F, Zhang Z, Liu G (2013) Responses of natural runoff to recent climatic variations in the Yellow River basin, China. Hydrol Earth Syst Sci 17(11):4471–4480. https://doi.org/10.5194/hess-17-4471-2013

Tang X, Fu G, Zhang S, Gao C, Wang G, Bao Z (2021) Attribution of climate change and human activities to streamflow variations with a posterior distribution of hydrological simulations. Hydrol Earth Syst Sci Discuss, 1–37. https://doi.org/10.5194/hess-2021-528

Tejedor A, Longjas A, Zaliapin I, Foufoula-Georgiou E (2014) Transport and Vulnerability in River Deltas: a graph-theoretic approach. arXiv preprint. https://arxiv.org/abs/1408.5834

Teng J, Chiew FHS, Vaze J, Marvanek S, Kirono DG (2012) Estimation of climate change impact on mean annual runoff across continental Australia using Budyko and Fu equations and hydrological models. Journal Hydrometeorol 13(3):1094–1106. https://doi.org/10.1175/JHM-D-11-097.1

Thavhana MP, Savage MJ, Moeletsi ME (2018) SWAT model uncertainty analysis, calibration and validation for runoff simulation in the Luvuvhu River catchment, South Africa. Phys Chem Earth, Parts a/b/c 105:115–124. https://doi.org/10.1016/J.PCE.2018.03.012

Thieme M, Birnie-Gauvin K, Opperman JJ, Franklin PA, Richter H, Baumgartner L, Ning N, Vu AV et al (2023) Measures to safeguard and restore river connectivity. Environ Rev. https://doi.org/10.1139/er-2023-0019

Tian H, Lu C, Chen G, Xu X, Liu M, Ren W, Tao B, Sun G, Pan S, Liu J (2011) Climate and land use controls over terrestrial water use efficiency in monsoon Asia 340:322–340. https://doi.org/10.1002/eco

Tirkey AS, Ghosh M, Pandey AC, Shekhar S (2018) Assessment of climate extremes and its long-term spatial variability over the Jharkhand state of India. Egypt J Remote Sens Space Sci 21(1):49–63. https://doi.org/10.1016/j.ejrs.2016.12.007

Tomer MD, Schilling KE (2009) A simple approach to distinguish land-use and climate-change effects on watershed hydrology. J Hydrol 376(1–2):24–33. https://doi.org/10.1016/J.JHYDROL.2009.07.029

Tongal H, Booij MJ (2018) Simulation and forecasting of streamflows using machine learning models coupled with base flow separation. J Hydrol 564:266–282. https://doi.org/10.1016/j.jhydrol.2018.07.004

Troin M, Vallet-Coulomb C, Piovano E, Sylvestre F (2012) Rainfall-runoff modeling of recent hydroclimatic change in a subtropical lake catchment: Laguna Mar Chiquita, Argentina. J Hydrol 475:379–391. https://doi.org/10.1016/j.jhydrol.2012.10.010

Tuteja NK, Vaze J, Teng J, Mutendeudzi M (2007) Partitioning the effects of pine plantations and climate variability on runoff from a large catchment in southeastern Australia. Water Resour Res 43(8):8415. https://doi.org/10.1029/2006WR005016

U.S. Fish and Wildlife Service (2011) Restoration of the Elwha River ecosystem. Archived from the original (PDF) on 1 October 2017

Uniyal B, Jha MK, Verma AK (2015) Assessing climate change impact on water balance components of a River basin using SWAT model. Water Resour Manage 29(13):4767–4785. https://doi.org/10.1007/s11269-015-1089-5

USGS Earth Explorer (2022) Landsat TM and OLI Image, United States Geological Survey Earth Explorer. https://earthexplorer.usgs.gov/. Accessed 06.10.2022

Verma N, Dash P (2018) Hydrologic response to land use change and climate variability in an ungauged basin, North-Western Himalaya, India. Analele Stiintifice Ale Universitatii" Alexandru Ioan Cuza" Din Iasi-Seria Geografie 63(1):54–82

Visakh S, Raju PV, Kulkarni SS, Diwakar PG (2019) Inter-comparison of water balance components of river basins draining into selected delta districts of Eastern India. Sci Total Environ 654:1258–1269. https://doi.org/10.1016/j.scitotenv.2018.11.162

Visakh S, Raju PV (2018) Long term trends in intensity and distribution of hydrological components of Brahmani and Baitarani basin using SWAT hydrological model. In: International soil and water assessment tool conference at IIT Madras

Vörösmarty CJ, Green P, Salisbury J, Lammers RB (2000) Global water resources: vulnerability from climate change and population growth. Science 289(5477):284–288. https://doi.org/10.1126/science.289.5477.284

Wang D, Cai X, (2010) Recession slope curve analysis under human interferences. Adv Water Resour 33:1053–1063

Wang K, Wang P, Liu J, Sparrow M, Haginoya S, Zhou X (2005) Variation of surface albedo and soil thermal parameters with soil moisture content at a semi-desert site on the western Tibetan Plateau. Bound-Layer Meteorol 116(1):117–129. https://doi.org/10.1007/S10546-004-7403-Z

Wang H, Yang Z, Saito Y, Liu JP, Sun X (2007) Stepwise decreases of the Huanghe (Yellow River) sediment load (1950–2005): impacts of climate change and human activities. Global Planet Change 57(3–4):331–354. https://doi.org/10.1016/j.gloplacha.2007.01.003

Wang J, Hong Y, Gourley J, Adhikari P, Li L, Su F (2010) Quantitative assessment of climate change and human impacts on long-term hydrologic response: a case study in a sub-basin of the Yellow River, China. Int J Climatol 30(14):2130–2137. https://doi.org/10.1002/JOC.2023

Wang J, Ishidaira H, Xu ZX (2012a) Effects of climate change and human activities on inflow into the Hoabinh Reservoir in the Red River basin. Procedia Environ Sci 13:1688–1698. https://doi.org/10.1016/J.PROENV.2012.01.162

Wang S, Yan M, Yan Y, Shi C, He L (2012b) Contributions of climate change and human activities to the changes in runoff increment in different sections of the Yellow River. Quatern Int 282:66–77. https://doi.org/10.1016/j.quaint.2012.07.011

Wang GQ, Zhang JY, Pagano TC, Lin JL, Liu CS (2013a) Identifying contributions of climate change and human activity to changes in runoff using epoch detection and hydrologic simulation. J Hydrol Eng 18(11):1385–1392. https://doi.org/10.1061/(asce)he.1943-5584.0000559

Wang W, Shao Q, Yang T, Peng S, Xing W, Sun F, Luo Y (2013b) Quantitative assessment of the impact of climate variability and human activities on runoff changes: a case study in four catchments of the Haihe River basin, China. Hydrol Process 27(8):1158–1174. https://doi.org/10.1002/HYP.9299

Wang G, Zhang J, Li X, Bao Z, Liu Y, Liu C, He R, Luo J (2017) Investigating causes of changes in runoff using hydrological simulation approach. Appl Water Sci 7(5):2245–2253. https://doi.org/10.1007/S13201-016-0396-1/TABLES/2

Wang J, Gao Y, Wang S (2018) Assessing the response of runoff to climate change and human activities for a typical basin in the Northern Taihang Mountain. China. Journal of Earth System Science 127(3):1–15. https://doi.org/10.1007/s12040-018-0932-5

Wang X et al (2022) Effect of river damming on nutrient transport and transformation and its countermeasures. Front Mar Sci 9:1078216. https://doi.org/10.3389/fmars.2022.1078216

Wang D, Hejazi M (2011) Quantifying the relative contribution of the climate and direct human impacts on mean annual streamflow in the contiguous United States. Water Resour Res 47(9). https://doi.org/10.1029/2010WR010283

Wang G, Xia J, Che J (2009a) Quantification of effects of climate variations and human activities on runoff by a monthly water balance model: a case study of the Chaobai River basin in northern China. Water Resour Res 45(7):0–11. https://doi.org/10.1029/2007WR006768

Wang G, Zhang J, He R, Jiang N, Jing X (2009b) Runoff reduction due to environmental changes in the Sanchuanhe River basin 23(2):174–180

Wang C, Wang S, Fu B, Zhang L (2016) Advances in hydrological modelling with the Budyko framework 40(3):409–430. https://doi.org/10.1177/0309133315620997

Wang M, Shao Y, Jiang Q, Xiao L, Yan H, Gao X, Wang L, Liu P (2020) Impacts of climate change and human activity on the runoff changes in the Guishui River Basin. Land 9(9). https://doi.org/10.3390/LAND9090291

Wang S, Li Q, Wang J (2021) Quantifying the contributions of climate change and human activities to the dramatic reduction in runoff in the taihang mountain region, china. Appl Ecol Environ Res 19(1):119–131. https://doi.org/10.15666/aeer/1901_119131

Wei X, Zhang M (2010) Quantifying streamflow change caused by forest disturbance at a large spatial scale: a single watershed study. Water Resour Res 46(12). https://doi.org/10.1029/2010WR009250

Wigmosta MS, Lettenmaier DP (1999) A comparison of simplified methods for routing topographically-driven subsurface flow. Water Resour Res 35:255–264

Wijngaard JB, Klein Tank AMG, Können GP (2003) Homogeneity of 20th century European daily temperature and precipitation series. Int J Climatol 23(6):679–692. https://doi.org/10.1002/joc.906

Williams JR (1995) The EPIC model. In: Singh VP (ed) Computer models of watershed hydrology, Chapter 25, Water Resources Publications, Highlands Ranch, pp 909–1000

Winton RS, Calamita E, Wehrli B (2019) Reviews and syntheses: dams, water quality and tropical reservoir stratification. Biogeosciences 16:1657–1671. https://doi.org/10.5194/bg-16-1657-2019

Wohl E et al (2021) A review of the impacts of dams on the hydromorphology of tropical rivers. Sci Total Environ 794:148686. https://doi.org/10.1016/j.scitotenv.2021.148686

Wolf S, Esser V, Lehmkuhl F et al (2022) Long-time impact of a large dam on its downstream river's morphology: determined by sediment characteristics, pollutants as a marker, and numerical modelling. J Sediment Environ 7:403–424. https://doi.org/10.1007/s43217-022-00103-9

Wu X, Li HE, Dong Y, Lin QC (2011) Calculation of ecological basic flow of Weihe River based on basic flow ratio method. Transactions of the CSAE 27(10):154–159. https://doi.org/10.3969/j.issn.1002-6819.2011.10.027

Wu J, Miao C, Yang T, Duan Q, Zhang X (2018) Modeling streamflow and sediment responses to climate change and human activities in the Yanhe River, China. Hydrol Res 49(1):150–162. https://doi.org/10.2166/nh.2017.168

Wu CS, Yang SL, Lei Y (2012) Quantifying the anthropogenic and climatic impacts on water discharge and sediment load in the Pearl River (Zhujiang), China (1954–2009). J Hydrol 452–453:190–204. https://doi.org/10.1016/j.jhydrol.2012.05.064

Wu J, Zheng H, Xi Y (2019) SWAT-based runoff simulation and runoff responses to climate change in the headwaters of the Yellow River, China. Atmosphere 10(9). https://doi.org/10.3390/atmos10090509

Xu J (2011) Variation in annual runoff of the Wudinghe River as influenced by climate change and human activity. Quatern Int 244(2):230–237. https://doi.org/10.1016/j.quaint.2010.09.014

Xu A (2025) The Indian River Delta under the Interweaving of climate change and human activities: multidimensional impacts from waterlogging to salinization. Theor Natural Sci 102:69–74

Xu X, Yang H, Yang D, Ma H (2013) Assessing the impacts of climate variability and human activities on annual runoff in the Luan River basin, China. Hydrol Res 44(5):940–952. https://doi.org/10.2166/nh.2013.144

Xu X, Yang D, Yang H, Lei H (2014a) Attribution analysis based on the Budyko hypothesis for detecting the dominant cause of runoff decline in Haihe basin. J Hydrol 510:530–540. https://doi.org/10.1016/J.JHYDROL.2013.12.052

Xu X, Yang D, Yang H, Lei H (2014b) Attribution analysis based on the Budyko hypothesis for detecting the dominant cause of runoff decline in Haihe basin. J Hydrol 510:530–540. https://doi.org/10.1016/j.jhydrol.2013.12.052

Xue L, Yang F, Yang C, Chen X, Zhang L, Chi Y, Yang G (2017) Identification of potential impacts of climate change and anthropogenic activities on streamflow alterations in the Tarim River Basin, China. Sci Rep 7(1):1–12. https://doi.org/10.1038/s41598-017-09215-z

Yan Y, Xue B, Yinglan A, Sun W, Zhang H (2020) Quantification of climate change and land cover/use transition impacts on runoff variations in the upper Hailar Basin, NE China. Hydrol Res 51(5):976–993. https://doi.org/10.2166/nh.2020.022

Yan T, Bai J, Lee Zhi Yi A, Shen Z (2018) SWAT-simulated streamflow responses to climate variability and human activities in the Miyun Reservoir basin by considering streamflow components. Sustainability (Switzerland) 10(4). https://doi.org/10.3390/su10040941

Yang L, Feng Q, Yin Z, Deo RC, Wen X, Si J, Li C (2017) Separation of the climatic and land cover impacts on the flow regime changes in two watersheds of Northeastern Tibetan Plateau. Adv Meteorol 2017. https://doi.org/10.1155/2017/6310401

Yang Q, Luo S, Wu H, Wang G, Han D, Lü H, Shao J (2019) Attribution analysis for runoff change on multiple scales in a humid subtropical basin dominated by forest, East China. Forests 10(2). https://doi.org/10.3390/f10020184

Yang Y, Tian F (2009) Abrupt change of runoff and its major driving factors in Haihe River Catchment, China. J Hydrol 374(3–4):373–383. https://doi.org/10.1016/J.JHYDROL.2009.06.040

Yang H, Yang D (2011) Derivation of climate elasticity of runoff to assess the effects of climate change on annual runoff. Water Resour Res 47:W07526. https://doi.org/10.1029/2010WR009287

Yang H, Yang D, Hu Q (2014) An error analysis of the Budyko hypothesis for assessing the contribution of climate change to runoff. 5375–5377. 10.3969

Yang S, Zhang Z, Kousky VE, Higgins RW, Yoo S-H, Liang J, Fan Y (2008) Simulations and seasonal prediction of the Asian summer monsoon in the NCEP Climate Forecast System. J Clim 21(15):3755–3775. https://doi.org/10.1175/2008JCLI1961.1

Ye X, Zhang Q, Liu J, Li X, Xu CY (2013) Distinguishing the relative impacts of climate change and human activities on variation of streamflow in the Poyang Lake catchment, China. J Hydrol 494:83–95. https://doi.org/10.1016/J.JHYDROL.2013.04.036

Yin J, He F, Jiu Xiong Y, Yu Qiu G (2017) Effects of land use/land cover and climate changes on surface runoff in a semi-humid and semi-arid transition zone in northwest China. Hydrol Earth Syst Sci 21(1):183–196. https://doi.org/10.5194/hess-21-183-2017

Yozgatligil C, Yazici C (2016) Comparison of homogeneity tests for temperature using a simulation study. Int J Climatol 36(1):62–81. https://doi.org/10.1002/joc.4329

Yuan Z, Xu J, Meng X, Wang Y, Yan B, Hong X (2019) Impact of climate variability on blue and green water flows in the Erhai Lake Basin of Southwest China. Water (Switzerland) 11(3). https://doi.org/10.3390/w11030424

Zealand MC, Burn HD, Simonovic PS (1999) Streamflow forecasting using artificial neural network. Water Energy Int 57(1):30–37

Zeng F, Ma MG, Di DR, Shi WY (2020) Separating the impacts of climate change and human activities on runoff: a review of method and application. Water (Switzerland) 12(8):2201. https://doi.org/10.3390/w12082201

Zeng S, Zhan C, Sun F, Du H, Wang F (2015) Effects of climate change and human activities on surface runoff in the Luan River Basin. Adv Meteorol 2015:1–12. https://doi.org/10.1155/2015/740239

Zeng S, Xia J, Du H (2013) Separating the effects of climate change and human activities on runoff over different time scales in the Zhang River basin. Stoch Environ Res Risk Assess 28(2):401–413. https://doi.org/10.1007/S00477-013-0760-8

Zhan C, Niu C, Song X, Xu C (2013) The impacts of climate variability and human activities on streamflow in Bai River basin, northern China. Hydrol Res 44(5):875–885. https://doi.org/10.2166/NH.2012.146

Zhang X, Zhang L, Zhao J, Rustomji P, Hairsine P (2008) Responses of streamflow to changes in climate and land use/cover in the Loess Plateau, China. Water Resour Res 44(7):1–12. https://doi.org/10.1029/2007WR006711

Zhang Y, Guan D, Jin C, Wang A, Wu J, Yuan F (2011) Analysis of impacts of climate variability and human activity on streamflow for a river basin in northeast China. J Hydrol 410(3–4):239–247. https://doi.org/10.1016/j.jhydrol.2011.09.023

Zhang A, Zhang C, Fu G, Wang B, Bao Z, Zheng H (2012) Assessments of impacts of climate change and human activities on runoff with swat for the Huifa River Basin, Northeast China. Water Resour Manag 26(8):2199–2217. https://doi.org/10.1007/S11269-012-0010-8

Zhang Y, Xia J, Shao Q, Zhai X (2013) Water quantity and quality simulation by improved SWAT in highly regulated Huai River Basin of China. Stoch Env Res Risk Assess 27(1):11–27. https://doi.org/10.1007/s00477-011-0546-9

Zhang C, Zhang B, Li W, Liu M (2014a) Response of streamflow to climate change and human activity in Xitiaoxi river basin in China. Hydrol Process 28(1):43–50. https://doi.org/10.1002/hyp.9539

Zhang X, Dong Z, Gupta H, Wu G, Li D (2016) Impact of the Three Gorges Dam on the hydrology and ecology of the Yangtze River. Water 8(12):590. https://doi.org/10.3390/w8120590

Zhang L, Karthikeyan R, Bai Z, Srinivasan R (2017a) Analysis of streamflow responses to climate variability and land use change in the Loess Plateau region of China. CATENA 154:1–11. https://doi.org/10.1016/j.catena.2017.02.012

Zhang Y, Zhong PA, Chen J, Bing J, Xu D, Wang M (2017b) Impacts of climate change and human activities on the three Gorges Reservoir inflow. Water (Switzerland) 9(12):1–14. https://doi.org/10.3390/w9120957

Zhang L, Nan Z, Yu W, Zhao Y, Xu Y (2018a) Comparison of baseline period choices for separating climate and land use/land cover change impacts on watershed hydrology using distributed hydrological models. Sci Total Environ 622–623:1016–1028. https://doi.org/10.1016/j.scitotenv.2017.12.055

Zhang H, Meng C, Wang Y, Wang Y, Li M (2020a) Comprehensive evaluation of the effects of climate change and land use and land cover change variables on runoff and sediment discharge. Sci Total Environ 702:134401. https://doi.org/10.1016/J.SCITOTENV.2019.134401

Zhang L, Wang C, Liang G, Cui Y, Zhang Q (2020b) Influence of land use change on hydrological cycle: application of SWAT to Su-Mi-Huai area in Beijing, China. Water (Switzerland) 12(11):1–17. https://doi.org/10.3390/w12113164

Zhang Y, Wu X, Wu S, Dai J, Yu L, Xue W, Wang F, Gao A, Xue C (2021) A framework for methodological options to assess climatic and anthropogenic influences on streamflow. Front Environ Sci 9(October):1–12. https://doi.org/10.3389/fenvs.2021.765227

Zhang X, Fang C, Wang Y, Lou X, Su Y, Huang D (2022) Review of effects of dam construction on the ecosystems of river estuary and nearby marine areas. Sustainability 14:5974. https://doi.org/10.3390/su14105974

Zhang L, Podlasly C, Ren Y, Feger K, Wang Y, Schwärzel K (2014) Separating the effects of changes in land management and climatic conditions on long-term stream fl ow trends analyzed for a small catchment in the Loess Plateau region, NW China 1293(2013):1284–1293. https://doi.org/10.1002/hyp.9663

Zhang T, Wang Y, Wang B, Feng P (2018) Understanding the main causes of runoffchange by hydrological modeling: a case study in Luanhe river basin, North China. Water (Switzerland) 10(8). https://doi.org/10.3390/w10081028

Zhao G, Tian P, Mu X, Jiao J, Wang F, Gao P (2014) Quantifying the impact of climate variability and human activities on streamflow in the middle reaches of the Yellow River basin, China. J Hydrol 519:387–398. https://doi.org/10.1016/j.jhydrol.2014.07.014

Zhao A, Zhu X, Liu X, Pan Y, Zuo D (2016) Impacts of land use change and climate variability on green and blue water resources in the Weihe River Basin of northwest China. CATENA 137:318–327. https://doi.org/10.1016/j.catena.2015.09.018

Zhao F, Xu Z, Zhang L, Zuo D (2009) Streamflow response to climate variability and human activities in the upper catchment of the Yellow River Basin. Sci China Ser E: Technol Sci 52(11):3249–3256. https://doi.org/10.1007/S11431-009-0354-3

Zhao F, Zhang L, Xu Z, Scott DF (2010) Evaluation of methods for estimating the effects of vegetation change and climate variability on streamflow. Water Resour Res 46(3). https://doi.org/10.1029/2009WR007702

Zheng H, Zhang L, Zhu R, Liu C, Sato Y, Fukushima Y (2009) Responses of streamflow to climate and land surface change in the headwaters of the Yellow River Basin 45:1–9. https://doi.org/10.1029/2007WR006665

Zhou S, Zhang W, Guo Y (2020) Impacts of climate and land-use changes on the hydrological processes in the amur river basin. Water (Switzerland) 12(1):22–25. https://doi.org/10.3390/w12010076

Zhou Y, Lai C, Wang Z, Chen X, Zeng Z, Chen J, Bai X (2018) Quantitative evaluation of the impact of climate change and human activity on runoff change in the Dongjiang River Basin, China. Water (Switzerland) 10(5). https://doi.org/10.3390/w10050571

Zhou Y, Yang Z, Zhang D, Jin X, Zhang J (2015) Intercatchment comparison of flow regime between the Hailiutu and Huangfuchuan rivers in the semi-arid Erdos Plateau, Northwest China. Hydrolog Sci J 60:688–705

Zuo D, Xu Z, Wu W, Zhao J, Zhao F (2014) Identification of streamflow response to climate change and human activities in the wei river Basin, China. Water Resour Manag 28(3):833–851. https://doi.org/10.1007/s11269-014-0519-0

Zuo D, Xu Z, Yao W, Jin S, Xiao P, Ran D (2015) Assessing the effects of changes in land use and climate on runoff and sediment yields from a watershed in the Loess Plateau of China. Sci Total Environ 544:238–250. https://doi.org/10.1016/j.scitotenv.2015.11.060

Index

S. Mondal and R. K. Paul, *Climate Variability and Human Interference*, Deltas of the World, https://doi.org/10.1007/978-3-032-13257-4

MIX
Papier aus verantwortungsvollen Quellen
Paper from responsible sources
FSC® C105338

If you have any concerns about our products,
you can contact us on
ProductSafety@springernature.com

In case Publisher is established outside the EU,
the EU authorized representative is:
Springer Nature Customer Service Center GmbH
Europaplatz 3, 69115 Heidelberg, Germany

Printed by Libri Plureos GmbH
in Hamburg, Germany